The
BLACK RHINOS
of NAMIBIA

ALSO BY RICK BASS

The Deer Pasture

Wild to the Heart

The Watch

Oil Notes

Winter

The Ninemile Wolves

Platte River

In the Loyal Mountains

The Lost Grizzlies

The Book of Yaak

The Sky, the Stars, the Wilderness

Where the Sea Used to Be

Fiber

The New Wolves

Brown Dog of the Yaak

Colter

The Hermit's Story

The Roadless Yaak *(editor)*

Caribou Rising

Falling from Grace in Texas *(co-editor, with Paul Christensen)*

The Diezmo

The Lives of Rocks

Why I Came West

The Wild Marsh

The Blue Horse

Nashville Chrome

The Heart of the Monster *(with David James Duncan)*

The Heart Beneath the Heart

In My Home There Is No More Sorrow

A Thousand Deer

All the Land to Hold Us

The

BLACK RHINOS
of NAMIBIA

Searching for Survivors in
the African Desert

RICK BASS

MARINER BOOKS
HOUGHTON MIFFLIN HARCOURT
Boston • New York

First Mariner Books edition 2013
Copyright © 2012 by Rick Bass

For information about permission to reproduce selections from this book,
write to Permissions, Houghton Mifflin Harcourt Publishing Company,
215 Park Avenue South, New York, New York 10003.

www.hmhbooks.com

Library of Congress Cataloging-in-Publication Data
Bass, Rick, date.
The black rhinos of Namibia : searching for survivors in the
African desert / Rick Bass.
p. cm.
ISBN 978-0-547-05521-3   ISBN 978-0-544-00233-3 (pbk.)
1. Black rhinoceros—Namibia. 2. Bass, Rick, date.—Travel—Namibia.
3. Namibia—Description and travel. 4. Wildlife conservation. I. Title.
QL737.U63B37 2012
599.66'8—dc23
2011051598

Book design by Melissa Lotfy
Map by Jacques Chazaud

DOC 10 9 8 7 6 5 4 3 2 1

Contents

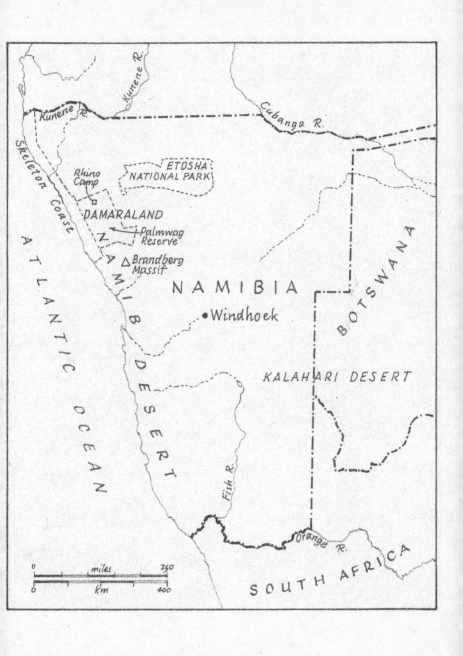

KUNENE R.

Kunene R.

Cubanga R.

Skeleton Coast

Rhino Camp

ETOSHA NATIONAL PARK

DAMARALAND

Palmwag Reserve

△ Brandberg Massif

NAMIB DESERT

NAMIBIA

• Windhoek

BOTSWANA

ATLANTIC OCEAN

KALAHARI DESERT

Fish R.

Orange R.

SOUTH AFRICA

miles
0   250
km
0   400

Prologue

I had been apprehensive about traveling to Africa, not yet understanding, as I do now, that the world *is* Africa: that Africa has been at the back of the world's curve for so long that it is now nearing the front again; that the rest of the world, which came from Africa, is becoming Africa again, as if the secret yearnings of an older, more original world are beginning to stir once more, desiring and now seeking reunification by whatever means possible—perhaps subtly, or perhaps in a grandiose way.

There is less and less a line, invisible or otherwise, between Africa and the world. And rather than arousing alarm—or is this my imagination?—it seems possible that as Africa's long woes and experiences become increasingly familiar to the larger world —radiating, as the origin and then expansion of certain species, including our own, is said to have radiated from Africa, into the larger or farther and newer world—we are turning to Africa not with quite so much colonial patronizing, but with greater respect, partnership.

There are those elsewhere in the world recognizing now that although Africa cannot by certain measurements be said to have prospered, it has, after all, survived—while many in the United States, for instance, exponentially less tested, are already buckling and fragmenting, falling apart at the seams. I am not saying our country yet has a whiff or taste of Africa's troubles—but I am suggesting that perhaps our own little sag is creating a space within us for something other than arrogance, and maybe even something other than inattention.

One country in Africa, Namibia, is fixing one problem—and I will not label it a small, medium, or large problem—with creativity and resolve. That's one problem solved, with a near eternity of problems still remaining. But it's a start.

We in the United States, on the other hand, are moving backwards: removing nothing from our checklist of either social or environmental woes—still proceeding, with the absurd premise that there is a wall between the two—and, in fact, adding to our lengthy checklist of unsolved problems and crises. Often we create new problems as we go, trudging into the next century with considerable unease, as if not only poorly sighted but possessing none of the other sensors at all, compassion included. Moving forward into the twenty-first century, but backwards into time and history, while some countries in Africa (and elsewhere) inch forward.

What is the individual's duty in a time of war—ecological and otherwise?

What is the individual's duty in a time of world war?

Always, the two most time-tested answers seem to arise: to bear witness, and to love the world more fully and in the moment, as it becomes increasingly suspect that future such moments will be compromised, or perhaps nonexistent.

And yet: one would be a fool to come away silently from the Namib Desert, having seen what I've seen—people in a nearly waterless land continuing to dream and try new solutions that are land- and community-based, and who move forward with pride and vigor and, perhaps rarest and most valuable of all these days, the vitality of hope.

The rhino—guardian of this hard edge of the world, pushed here to the precipice—is giving them hope.

Part I

PASTORAL

HAVING ALWAYS BEEN SOMEWHAT CLUMSY IN THE WORLD, and growing more so as I begin to age, I agreed to travel to Africa with my friend Dennis with some apprehension. Dennis, a burly fellow whose presence in the world—having had his arm nearly whacked off at the shoulder by a float plane propeller, and having been charged and knocked down by grizzly bears, enthusiastic rugby players, and others—is still, even after a half century, sometimes extremely exuberant. He tends to see only the positive lights of the world—the bounty over the next rise—whereas I am a practical worrier. And knowing of my clumsiness, I worried that I might make mistakes—simple errors in local customs, out in the bush—that would conspire then to be our undoing. I wasn't so concerned for myself, but was keenly aware of my responsibilities as a parent, of the need to stick around for my girls.

And yet: I wanted to see a rhino. And not just any old rhino. The white rhinos of South Africa were at that time prospering, inhabiting the brush and veldt country, gigantic and mythic creatures whose appearance, sudden or otherwise, amid the leafy, thorny scrub, or seen grazing at dawn on the pastoral green of a bedewed meadow, should have pleased the desires of any middle-aged man beginning to wonder at what he might not yet have seen or known in the world.

But a white rhino evidently wasn't good enough. Dennis and the staff and students of his Round River Conservation Studies — a nonprofit group he founded about twenty years ago — were participating in a study of black rhinos. They are rarer and more estranged from the world, you could say, inhabiting the edge of the spooky and surreal Namib Desert, caught between the uninhabitable superheated giant sand dunes (some nearly two hundred feet high) that plunge down into the South Atlantic Ocean, and the scrabbling swell of human communities that cluster farther inland.

In this space between humankind and the uninhabitable abyss lived, and live, the last of the black rhinos, and the first of the black rhinos — the recolonizing stock, if the black rhinos are to ever be restored to the world they once strode in almost unimaginable numbers and with what must have once seemed like almost limitless distribution.

There is perhaps no greater animal that has been relegated and confined to so small and finite a space. Surely it would be an amazing sight to any traveler to witness, like a voyeur, the grace, elegance, and dignity with which these last rhinos inhabit the austere country that the world has bequeathed to them.

The Namib Desert is one of the oldest unchanged landscapes on earth. A meandering contour of basalt prairie that rests like the fuzzy light between dream and wakefulness, in this ribbon of land between the ocean's dunes and the last of the human communities — the out-flung, hardscrabble goat-herding villages — the rhino's desert, known informally in recent times as Damaraland, is almost identical, meteorologically and geologically, to how it was more than 130 million years ago.

Because it receives between only one and five inches of rain per year—year in and year out, across the eons—there is little vegetation that grows there, and, as with much of sub-Saharan Africa, life revolves, like a tiny model of our earth, around the presence of water. It is the nearly eternal absence of water that has shaped and sculpted everything in this part of the world—crafting each individual species, and then the movements of populations and cultures, and the relationships between these things, with such godlike intricacy and sophistication that it seems surely some foreknowledge must exist: for surely such intricacy of fit and design cannot be random or crafted on the fly, but instead was laid out earlier, as if by some ancient and celestial cartographer.

But these are middle-aged traveler's questions or musings, and surely not the black rhino's. I simply wanted to witness such a ponderous beast out upon such a naked and seemingly unsupportive landscape, and to see new things, and learn new things. I wanted to see the mesmerizing spill of wind-rounded basalt cobbles scattered to the horizon, each stone blood red and time-varnished with iridescent sheen similar to that of birds' intestines, nothing but Martian-red cobbles on that desert terrain for as far as the eye could see, and farther—for as far as the imagination could see.

Without quite understanding it at the time, I wanted to see also the face or at least a glimpse of the world-to-come—to witness the utter signature of environmental paucity—one of the largest species pushed to the maximum brink: the bitter edge that so many of our own species are hurtling toward here in the United States, and elsewhere in the world. How far away are

5

we, and so many of our endangered species—most notably, in my world back in Montana, the grizzly bear—from that same precipice?

Here, with the rhino, was a creature—the mythic made real —that, better than almost any other mammal, could tolerate a world of almost supernatural heat, and the extreme rarity of water. How many short years before we, even in the flush United States, might be entering such a future? Not just our grizzlies and our salmon, our cranes and darters and tanagers, but we-the-people, perhaps frailest and palest among the species? How far, that future?

I wanted to see the real creature, and I wanted to see the myth too. I wanted to see an animal so stolid in the world, even at the edge of apocalypse. I wanted to see what one journalist called "rhinos on the moon." I wanted to wake up or—perhaps the opposite—go to sleep, and enter another dream, and to then stand as close as possible to the edge of the dream of that other world, close enough to smell the dust and to bake in the dazzle of heat, and to hear the click of hoofs, and even the shifting of muscle.

I wanted to stand right at the edge of that world—no more than one step away, so that I might even choose to enter it— to witness the creatures that have perhaps lived far beyond their time, as if even the rhinos, having survived and flourished in their old world, have themselves come now to the edge of another dream, one that they too must choose to step into and pass through, and in so doing, be saved or lost.

If we cannot fully know the largest and most conspicuous things —the full nature of giant rhinos, for instance, or the future path of so studied a species as mankind, and our own relationship with the world—then how can we be expected to know or learn anything?

This I think is one of the things that attract the eye and the mind to the grand megafauna of the world. It is on the canvas of rhinos, blue whales, elephants, and grizzly bears that our increasingly benumbed eyes, and the rest of our stunned senses, besieged by the amperage of this era, can still seek order and understanding—can still see the master strokes, the primary strokes, of nature writ large. Certainly there is every bit as much beauty in the intermolecular structure, the crystalline lattices, of snowflakes, or in the frozen blood-crystals of a hibernating salamander. But the megafauna promise, sometimes, to make things simpler for us. A glimpse, a glance, reveals much: although even behind the veil of the obvious, or what seems obvious, there is surely much that remains hidden, even among giants.

Big animals, with the broad strokes of their movements and lives, can show us the world, and with those broad strokes lead us further into imagination. They can teach us how to consider small strokes as well. Sometimes when I ponder humans' place in the world, it seems to me that we are positioned eerily in the middle of almost all things, by which I do not mean the radiant center around which all of nature and order orbits, but rather, in the linear middle; not as dramatic as rhinos, for instance, but possessing (in some ways only) a bit more drama than, say, a snowshoe hare, a lemur, or a Norway rat.

We are well positioned in the world to act as sentries or witnesses from that midpoint and look back at the small while looking forward to the grand. And from that curious midpoint too we are able to see back into history, and through the tools of science, able to see some distance into the future.

So this was another of the concerns that I carried with me

to Namibia: that even at a species' bitter edge, there could be hope; and that as long as there is creativity and imagination, there can be hope. In my home valley, the Yaak, in northwest Montana, our own ultimate megafauna, the grizzly bear, has dwindled from perhaps three dozen, when I moved here nearly thirty years ago, to perhaps only one dozen. In the last seven years alone, there have been twenty-seven *known* human-caused mortalities of grizzlies in the Cabinet-Yaak ecosystem, the home of the most endangered population in North America. *Dwindled* isn't the right word; they're in freefall, and no public agency is stepping up to halt it. The activist Louisa Willcox and others have labeled the Cabinet-Yaak bears "the walking dead." These last surviving bears inhabit a landscape so dangerous to their continued survival—a national forest riddled by too many roads, too many clearcuts, and too many noxious weeds, and with no strategic plan in place to protect the low-elevation private lands that surround the Cabinet-Yaak country, and no planned corridors to connect this unique, lush landscape to other regional ecosystems—that even the last twelve living bears may now possess a punched ticket for extinction, unless dramatic measures are taken.

I could rant and rave about the absence of courage or conviction among government leaders under the feckless environmental guidance of George W. Bush; I could continue haranguing the agencies responsible for protecting this species, and this landscape, the way I and many others have been hectoring them for these past many years. But that hasn't gotten the grizzlies very far. In fact, on the contrary, we've been losing ground. Whatever it is that we've been doing, or not doing, the grizzlies can't stand much more of it. We—they—might have time for one more idea, one more game plan—*maybe*—and I think it will have to be a plan that takes into account the political cowardice, the degree of corporate ownership in this country, thick even into the halls of Congress and the White House.

To pretend otherwise in prescribing a cure—a real-world cure, as opposed to a pen-and-ink cure—would be like a physician looking at a patient riddled with cancer but then prescribing a course of treatment that does not acknowledge that cancer. It's taken me a long time to accept the belief that bitching and moaning is not going to be enough to get to the bastards. They are too insulated, too good at dodging and weaving and running out the clock. Even the temporary tonic of lawsuits and injunctions, while often life-saving for the grizzlies, are but stays of execution; even with those defenses, the Cabinet-Yaak grizzlies are hurtling toward extinction and could be completely gone in another ten years.

To save grizzlies in the West, we must find places where they, if not we, can avoid our consumptions and the countless clumsy mistakes we commit against the natural world each day. We must find a place where they are safe from our ceaseless hungers.

We also need to find a way to help the grizzly to be viewed locally as an asset, as well, rather than a liability: something whose increasing numbers, for instance, might trigger more economic assistance.

In my county—Lincoln County—our local unemployment and health issues aren't African in scope, but neither are they quite like anything else in America. Unemployment fluctuates between 12 and 25 percent, and one-third of the population tested has shown signs of pleural thickening in the lungs, a possible precursor to the fatal disease mesothelioma, courtesy of the W. R. Grace asbestos mine—once the largest in the world—which operated here up until 1990, despite the company and the government's knowing the toxic effects of their product and its harm to workers and townspeople.

The disease—Peter Grace's legacy, and the Reagan-Bush administration's legacy, takes years, even decades, to manifest itself, but has already killed more than two hundred in a small town of approximately 2,500, leading health care officials to speak of thus far having only encountered the tip of the iceberg.

Corporate avarice and manslaughter possess perhaps a tonal difference, and are different from the slave trade that was unleashed on Africa and the genocides that have marched almost uncontested against that continent, though in the end, to the victims, is not the result the same? Isn't dead dead?

And certainly, the AIDS pandemic—5.6 million people infected in South Africa alone—is a quantum universe away from the brutalities of mesothelioma, though again, only in scale, only in numbers. Here too, in Montana and the United States, Congress and the White House bicker in a way degrading to humanity over corporate responsibilities, over who should pay, and how much, if at all, for the treatments that can buy some wedges of time for the victims of mesothelioma, an utterly horrific disease.

The two stories, Africa's and my valley's, are not the same. And yet in traveling to Africa, I was surprised, upon arriving, to begin to suspect, as I looked around, that even coming from the backwoods of a forest far up in northwestern Montana, with a population of 150, I already had a pair of spectacles that would allow me to look at Africa.

Everything I would see would be new and different, but the prescription for my lenses did not need changing, and the stories were so eerily the same that despite the different and amazing cast of characters and elements, it seemed the only difference was one of time, not space; that the rhinos were the grizzlies, and Africa's Bantu our Sioux. It seemed that Story was all, that Story controlled everything. That the world was unfolding like the sweep of dominoes—whether by design or chance has long been argued, and probably will be argued for at least a while longer—or like the tops of grass gusting before a swirling wind that advances across the field so quickly that from a distance, one can see it all happening, the grass tops bending almost simultaneously, and yet one can see too the patterns of the wind's breath.

From such a perspective, the viewer can be mesmerized, spellbound by the grace of that vision, if relentlessness can be said to be a kind of grace.

To such a viewer, the wave of Europeans crashing onto the shores of America is surely little different from the wave of Europeans washing onto the shores of Africa's South Atlantic. The travelers, the voyagers, struggle ashore, still possessing, perhaps, remnants of the superior technology that allowed them to even attempt, or thus far succeed at, such a journey, where they are either attacked and vanquished by the natives, or where the natives instead aid, abet, and rescue the travelers, who in turn then subsume their rescuers, in one form or another—usually, but not always, violently—over the course of a year, a decade, a century: it matters not to the gust, the swirl, of wind. Stand on the timeline in any one place in America in 1776, or 1876 at the Little Bighorn, blink or nap for a moment, and one awakens, or reopens one's eyes to be standing in South Africa in 1976.

Closing one's eyes again, one advances twenty years, falls back fifty, skips ahead thirty, or even two hundred. It is all wind, and perhaps only the substrate below is firm—though a geologist knows that even this myth offers no truly enduring reassurance, for even the most durable and unchanged landscapes are dissolving and then being re-formed beneath that breath, worn smooth and eroded by wind and sun, ice and snow, and by the endless passage of life across the land, with the travelers etching tiny paths across the surface, and the forces of erosion focusing themselves then upon those trails, those downcuttings toward the source of the land itself, which, the geologist knows, is the fire, the fever, at the center of the earth.

A traveler leaving his or her small province is always surprised by this revelation—seeks and always finds the similarities, even in the midst of other differences, and is always surprised by them. And yet why should they surprise us—particularly the American traveler? What colonial arrogance still resides in each of us, like a virus—Bantu or Texan, New Yorker or Inuit—in some ways perhaps residing in us more strongly than ever, with each passing century, for us to think or assume that it, *the world,* should be any different here in Africa, or anywhere?

In some ways, I think it is our subconscious or unconscious dreams of landscapes—particularly a landscape not yet altered by humans—that lead us to believe that just because there may be an infinitude of possibilities on any one landscape, there can also be an infinitude of different stories.

We want our continents and their stories to be utterly different. In our innate hungers, we desire there to be a richness of stories, when our real wealth lies in the elements within that more common one-story. We forget what the geologist knows and does well to remember, which is the story of Pangaea. It is no Bible myth, no fevered dream of a shaman, that the earth was once whole, and the sea once total and complete; no pagan tale that the earth rose above the waters, and that the firmament was then spread and separated like clay, breaking apart and floating into different corners of this small planet.

City or country, mountain or desert, north or south, no matter: One of the ways we enter a new landscape or new environment is through story, and Dennis, who has only begun traveling to Namibia a few years ago, nevertheless has plenty of them. In the vacuum of my inexperience, his stories fall like rose petals, accumulating, twisting, seeking to collate into meaningful narrative.

Dennis doesn't vouch for the stories' veracity, but he tells me that others have told him that a camper will do well to sleep with a towel or blanket over his face for protection against hyenas. The hyenas, says Dennis—so he has heard—like to bite the face of sleeping victims, and so if they can't see the face, they will pass on by.

There's another story like that, he says, about lions, but he can't quite remember how it goes. It's something about lions under trees, he says: If you're out walking, and happen to see a lion under a tree, don't look at the lion, or the lion will be obligated to eat you. Or maybe it's that way only for sleeping lions: If the lion sees you seeing *it* sleeping under the tree, it has to eat you. "Anyway," Dennis says, "whatever it is, if you walk up on a lion under a tree, pretend you didn't see it."

He didn't really know whether to believe the stories about the black mambas or not. "Everyone said not to drive with your windows down, or else the mambas would jump up off the road and come in through the window. They lie on the side of the road, looking like sticks, and then when a car or truck goes past, they jump up." Dennis shrugs. "We had a student who had one jump up against the window of her Land Cruiser," he says. "She had the window rolled halfway up to keep some of the dust out, and she saw what she thought was a stick lying on the road, but when she went past, it rose up on its tail and bit at the glass," he says. "If the window hadn't been rolled halfway up . . ."

It takes a long time to learn to fit in the world. In America, we don't so much control nature as simply run away from it. We seek to neuter nature, but really end up numbing or sterilizing only ourselves.

Bleary from the seemingly relentless travel—Yaak to Spokane to Salt Lake to Atlanta to Cape Town to Windhoek, forty-eight hours of nonstop travel, though only a hundred years ago the journey would have taken weeks if not months—we overnight at a country club, where there is a garish casino, an outdoor pool, all-night sprinklers hissing to keep a desert lawn green, a vast buffet, and, seated on couches by the bar, ladies for hire, dressed as garishly as reptiles, with beaded sequined dresses, frosted hair, supernaturally long eyelashes, and makeup like war paint. Boots up to their knees. One of them is wearing what appears to be a bulletproof metallic vest. They are waiting for their dates, rumpled-looking businessmen still in their suits, and from the dining room below, I can hear a recording of Bing Crosby singing "I'll Be Home for Christmas."

Is this, or is this not, Africa? You would think that the older a person gets—the more experience a traveler obtains—the less that homesickness would be a problem for him or her. But I have not found that to be the case for myself. It grows stronger, and comes more easily, comes quicker, with each passing year. For a homebody such as myself, any venture into the outside world can be like going out onto thin ice, with my serenity in the world able to be disrupted by a single sound, a single unpleasant event: as if some saturation point has been reached wherein the heart can be made heavy by even one tiny, dangerous thing. Like the last straw placed upon the camel's back, even a trip to town, some days, can elicit such a response—*I just want to be home*—and here, even while on the trip of a lifetime, I can find myself trembling suddenly, as if about to fall through that ice, wanting nothing more than to be going about the mundane tasks of everyday life, and of parenting. Here in Windhoek, as I'm listening to Bing Crosby, and weary from the travel, the homesickness comes creeping in like a plague, or like floodwaters, and there's nothing I can do about it but ride it out.

It does pass—I've learned that, in my travels—but when I'm in the midst of it, the only three things that can help are to get lost in a good book, or to spend myself physically—to walk, or run, or swim—or, sometimes, if I'm lucky, to sleep it off, as one might go to sleep with an injury or other illness, hoping that the next day—miraculously—might bring healing, and the absence of pain.

What a delicate aesthete I am for the gearing of my interior world to be so easily rocked, upset by a single, sad image— garish prostitutes in a sea of AIDS, electronic Christmas decorations from home, the fruitless ambition and ceaseless waste of the sprinklers.

Why fight for the rhinos, the grizzlies, and the glaciers, with the world being turned now so violently upside down?

"Doesn't everything die at last, and too soon?" writes the poet Mary Oliver.

We pick up our rental vehicle the next morning. The proprietor hands us a detailed map of Namibia and shows us the roads we can drive on in which our insurance will be in effect. It is only a handful of roads—the rest of the country is presumably gravel and sand dunes—and while he doesn't forbid us to drive there, he does tell us that we'll be completely liable for the vehicle—a Toyota Land Cruiser—which is running a cool $60K in Windhoek these days. Dennis and I look at each other and then Dennis says, "It's best not to think about it," and signs the contract.

The truck's best feature is the fold-out tents welded to a platform on top. Each tent has an aluminum base that folds out (like a sandwich being opened, I imagine, thinking of the lions), with the tent itself then appearing, rising like a magical pop-up toy. (Or, like a beacon that dinner is served, the cathedral-like tent appearing in silhouette against the reddening sky of dusk, and the lions coughing, roaring to one another, padding silently toward the canvas skyscraper on dagger-clawed paws wider than a man's outstretched hands, breathing their hot and anticipatory breath through a cage of teeth, many of which are as thick in diameter as a man's finger . . .)

"We'll take it," Dennis says. The traffic of downtown Windhoek is glinting, flowing around us like bloodstream amoebae, cars are honking and the heat is building up to its full oven intensity, and I realize that short of a tank, we are not going to find a vehicle that will keep us safe against everything. Rhinos and elephants can tip us over, lions and hyenas can leap spryly onto the top of the vehicle, mambas can sail through the air like arrows unleashed by malignant archers; and that if complete separation from those dangers, however slight, is my desire, I have come to the wrong country. Here, there is still no full prophylactic against the larger and older world.

We drive north in our fancy new rig with its fancy new tires. On his last trip to Africa, up into Botswana, Dennis went through ten flat tires, with service stations sometimes hundreds of miles apart. We have only two spares with us on this journey, but again, what are we to do: stuff ten spares, wheels and all, into the back, along with all our cooking and camping gear?

And it's not exactly survival camping. We've got an extra five-gallon can of gas, a five-gallon jug of water, a sissy folding card table upon which to prepare and eat our meals, folding chairs, a gas stove, little aluminum pole ladders to use in ascending to our tents each evening, and, most amazing of all, a battery-operated refrigerator.

Still, softness aside, it's an adventure; it's all new. We muscle our way out of the centrifugal pull of the city, and soon the countryside reveals itself, the scrubby vegetation heat-burnished. Baboons saunter down the highway with the same assurance with which a human might walk the streets of his or her hometown. They look and act like humans—they look like humans who have escaped from a zoo, I think, or like the world remade—and they are the only thing out and about in the great heat, seeming unaffected by it: though we see others of them too, sitting atop the transformers on the poles for telephone and electrical lines, and still others sitting like old men in the shade beneath individual trees, the shapes and silhouettes of their bodies appearing exactly like those of men and women.

Except for the sauntering baboons, the country looks like south Texas—a little brushier, perhaps—and Dennis tells me that, just as in our country, fire suppression is upsetting the previous ecological labor of the centuries, upsetting the work of the eons. In central Namibia, Dennis and his students have been working with a cheetah conservation and rehabilitation program, Save the Cheetah, where one of the most common treatments for injured cheetahs involves eye surgery, because the cheetahs, used to opening up full-throttle on their native landscape, native grasslands, are now encountering too much brush and are tearing their eyes on thorns as they accelerate through the mazes of new brush, where for the tens of millions of years previous there was none.

Termite mounds begin to show up here and there, appearing through the patchwork of brush like pueblo chimneys, the beautiful brick-red color of sun-baked clay. How perfect the world is, how infinite and marvelous the adjustments! The story of Africa is the story of heat and paucity—heat, and the absence or conservation of moisture—and as they stipple the torturously hot land, the high towers of the termite mounds act as radiant supercoolers, providing a greater surface area with which to dump the land's heat, as if the earth here is a desperate, panting, even suffering animal.

The mounds aerate the soil below, too, in a land where otherwise an impermeable hardpan might bake itself into a life-robbing seal, separating the world above from the world below and cutting off the necessary transfer of carbon and other nutrients between the two. In this regard—visually as well as ecologically—the termite mounds are as dramatic as a tracheotomy, and we pass on, leaving the big city farther behind.

Now and again we encounter a town at the infrequent intersection of two roads, and I stare and gawk, trying to take in every little fragment of newness. The beer bottle caps pressed like artifacts into the heat-softened red clay roadsides, the warped hills upon which structures are built, the sagging power lines, the crowds and clots of people lingering back in the shade, the midday central activity around the grocery store, the blessed and complete absence of chain stores, and chain restaurants—there is no one turn of the key that says, *This is Namibia,* or none that I can see yet.

I don't know quite what I'm expecting. It's big and vast, but the country below the country is even more so; and I think that in part this is my instinctive fascination with rhinos, or even the idea of rhinos: that they are not so much metaphor as reality, in their enduring service as ambassadors between this world and that, mute witnesses and, for all we know, judges to the brief scald of human history across this land. Existing unchanged for so long —ten million years, at least, on the oldest unchanged landscape on earth—the Namib Desert has not so much as blinked, geologically speaking, in its last fifty-five million years, so it is as if the rhinos might always have been present, pre-known and pre-formed in that other world, and were awaiting only to be released; waiting to emerge, perhaps, as if resting just beneath some sun-baked hardpan, the erosion of which finally allowed the rhinos to step up and out and into the world.

It is the genius of beauty to make a thing look easy when in fact it has been anything but. In the end, looking at a rhino—examining and considering it closely—I wonder what else *could* have developed, out on this singular world?

Of course this desert-land had to sculpt a three-thousand-pound nearly blind racehorse with three-foot-long dagger horns, capable of eating poisonous plants and going without water for days. The fit is astonishingly hard-gotten, with the story of evolution and natural selection a tale of such power and glory and, always, the finer twists of occasional mercy, that I sometimes wonder, no disrespect intended, how it would read had the Bible not given us the six-day short version.

Maybe this is one of the great and many differences between humankind and God, or a God: the former appreciating the play-by-play, fifty-million-years-in-the-making saga of *How to Make a Rhino*, while to the latter, such a feat might seem like only a quick magic trick, if that, a handkerchief waved over a top hat, with a surprise—a rabbit—appearing, much to the audience's surprise and entertainment.

Does the audience ask how the trick was done? *No.* Does the audience seek to understand each individual cell in the rabbit, and the effects of iodine transfer between and across cell walls, in the rabbit? *No.* The rabbit is summoned, the rabbit appears, and the audience gasps; the audience's attention is directed toward the magician.

And yet, would it not also be an act of worship, respect, and even obeisance to seek to examine and celebrate the magic, not just the magician? And would it not be a matter of great profanity to willfully erase—or even allow to be erased—some of the very creations upon which the magician labored longest?

Though I do not know it yet, this is one of the reasons I am in Namibia, and pursuing rhinos: to observe a story of recovery rather than continued loss and diminishment, and to witness the reemergence of not just any species, but one as magnificently dreamed and crafted as a rhino.

Would not the sight of such an animal reemerging into the world—climbing as if slowly back up the steps from the dusty basement in which it had long been imprisoned or contained—be, for all intents and purposes, like witnessing the animal's first emergence, first creation?

The tip of the first horn appearing first, perhaps, piercing the overburden of soil that lies across the land like an armored skin. The second horn appearing then, followed by the ears, and the top of the head.

A cloud of red dust, and a tremor in the earth as the animal labors to find purchase with its ponderous three-toed front feet. More shuddering and dust as the animal climbs all the way out then, perfectly formed and dreamed like some immense hatchling born from the earth itself; or born, perhaps, from the nature and essence of time, or the union of time and this open landscape: this one horribly austere landscape, so unlike any other.

The day seems to hang forever, with the flat, harsh light unchanging on the rocky, brushy landscape. We stop for groceries in Otjiwarongo, where the twists and braids of the baked bread are different. Lavishly designed rolls and muffins that would be the envy of any Parisian bistro or Pike Place bakery are piled high in handmade reed and wicker baskets. On the cold drink shelf, there are numerous Africa-made drinks—many of them are mango-based—though there are also some of the ubiquitous gold-green and cobalt blue "power" drinks such as I might find in any convenience store back home.

Still, the familiar is the exception—indeed, this one product and a certain red and white decorated carbonated beverage are the only two items with packaging I recognize—and even the act of navigating the narrow shopping aisles, with every single product, whether salt or olive oil or flour or sugar or raisins, as strange and dazzling as some unknown and never-before-seen species, is completely new.

A young man helps us carry our week's worth of groceries out to the truck. Earlier, another young man had come up to us in the store and introduced himself, asking our names and shaking our hands for no clear reason that I could discern; and though friendliness had been a part of it, there seemed to be a touch of something else to it. I thought I detected the urgency of commerce—that universal language—and then I felt bad for suspecting such a thing. Maybe he just wanted to try out his English on native speakers. It seemed more than that—perhaps, having marked us so obviously as tourists venturing into the bush, he wished to provide his services as a guide — if so, we would have to tell him our travel plans were already arranged — but again, maybe I was wrong and he was just being friendly.

His eyes had seemed almost desperate, however, and he had been reluctant to leave us — he was asking us where we lived, how we liked Africa, what animals we had seen, what our plans were, and so forth — and with our natural reticence or guardedness engaging more deeply with each new question, we finally were able to ease away and make a hopefully tactful transition back to the examination of cloves of garlic, a lemon, an orange. The freshest possible pork chops, the big brown eggs with the fewest cracks.

Back at the truck, as we're rearranging our gear, and sorting, stacking, and wedging the cold goods into the nifty little refrigerator, our friend from inside the store shows up beside us flashing a knife, a bright razor, and a handful of carved nuts. He's asking if we're married, asking if we have children, asking the names of our wives and children, asking if they would like some of his carvings.

Like a school of fish, his associates are surrounding us now, as if to pin us against the truck. Dennis sidles away to the right-hand driver's side, while I keep myself between the door and the marketeers.

The nuts are about half the size of a Ping-Pong ball and are carved with intricate scenes of African wildlife etched into the nut-brown skin to expose the cream color underneath. They dangle from leather thongs, looking, I suppose, like novel Christmas ornaments, and had I seen one in a store, I might have stopped to examine it briefly.

Unfamiliar, however, with this manner of shopping — defensive rather than offensive shopping—I am uninterested in the multitude of clacking nuts being waved at me, and am chagrined to see that my name has already been carved into at least one of them, as has Dennis's.

Somehow—even as I am in the act of saying goodbye, thank you, no thank you, and am climbing into the truck but rolling the window down for politeness—the conversation has become one of how much. Dennis is grinding the gears into reverse, and I am sorting through my pockets for the proper amount of buy-off money, counting out the beautiful bimetal gold and silver coins of various sizes and shapes, themselves festooned with various wildlife images—no dead presidents here—with some of the coins worn as smooth as river rocks, the once bright metal polished down to a duller sheen now not by the relentlessness of water or time but by the hands of humans, countless fingering transactions such as the one occurring even now, this brilliantly hot December day.

The coins are as beautiful as the nuts, and hence an even trade, but with tactful freedom a part of the deal now, and the Toyota clacking and grinding as Dennis backs quickly but carefully out of the parking lot, the throng falls back, parting and then recoalescing, sorting out the South African rand and Namibian dollars according to which vendor sold which nut, and the difference between our lives and affluence and privilege so vast as to seem cataclysmic, unfathomable, immoral. And yet, what can any one individual do to change the disorder and imbalance of the world? Were the situation reversed—were Dennis and I hawking nuts beside a rapidly retreating new rented Toyota—what would we wish for, what would we ask for, and what would be deemed moral?

We are in the world such a short period of time. What is the true path, or the best path, or the shining path? Where do its branches travel, day by day?

As we drive farther north and west, I brood, wishing I had bought more nuts: ashamed of the silly, reactive bargaining that came over me, and of my instinctive defensiveness with my money, when I had so much and they had so little.

It's crazy to be fretting about such a thing—a nut!—but it bothers me. Ironically—or maybe not ironically—one of the best things we from the pathologically affluent society of America can give is our support for the rhinos, and support of the rural communities that support the rhinos.

Rhinos, I tell myself, *not nuts. We're here for rhinos.* It's the old traveler's dilemma, one of the tourist's many indulgent rhetorical dilemmas—to buy or not to buy?

Fortunately, too, as it will turn out, we will have many subsequent opportunities to purchase additional nuts: at almost each and every stop.

For a long time, the light remains as it is, harsh and flat and glinting against an increasingly rugged landscape of red-rock basalt and scraggly brush, the vegetation surprisingly green, the result of the freshet of last week's rain—the rainy season, all one inch of it, having come and gone—and there seems now to be only one road in all of Namibia, and we are the only ones on it. We ascend rises and ridges, with more landscape appearing below us, and although as a geologist I know I am looking at some of the oldest earth there is, the effect is that it's all brand-new, rarely explored, and waiting to be explored, if not quite discovered.

The light is finally beginning to soften. We descend one long plateau and enter a narrow canyon that looks not unlike the red-rock vistas of Navajoland, complete with sand washes. In the umber light, we spy a herd of springbok, our first herd sighting of African ungulates. There are over a hundred different native species in Africa (North America has about one-tenth the number), and to the question of *Why?*—*Why so many?*—a scientist will likely answer *Grasses*—there are many grasses, and few trees, in such an arid land—and yet still, the numbers seem excessive, unless it is simply that ours are impoverished.

We're used to thinking of the Bible's creation story, the six-day run, as metaphor for world's beginning, and the so-called ascent of man—Day One the Precambrian, Day Two the single-celled organisms, Day Three the swimming vertebrates, Day Four the rise of reptiles and birds, and so on and so forth. But it occurs to me that in addition to being history, the story might also be prophecy: that indeed, if the story plays itself over and over, radiating into time and space like the ripples of a stone tossed in a pond, we might currently find ourselves in midstride of some further version—with those ripples and story replications expanding, I suppose, all the way out to the end of time, or the end of space, if such things exist. Or not; what does it matter?

Hence, in a larger story, a larger telling, the continent of Africa itself and all its history to date might be but Day One or Two—the dab of clay, the heated dust. And the fracture of Gondwanaland and the drifting across the sea was but Day Three—the oceans mixing with that clay and dust to make new continents, some with rumpled skin and others sere and plain.

What if we—in this present story, present scale—are only at Day Three or Four, or Day Five, tops?

Such a notion does not fit with anything we have been taught or told, or with the sometimes sinking feeling and other times urgency that seems to inform us that the world is full, that the story is running out of possibilities, that the end is very near, and that even time itself seems to be collapsing rather than expanding, ripple-like.

I certainly don't know. I do know that in Africa I felt even tinier against that scale of time than I do in Montana, standing on a windy ridge, tiny against the scale of space. And that either way, in both instances—atop the windy Montana ridge, or staring across the unchanged-in-time Namibian desert—it feels wonderful to understand more precisely how tiny we are.

The springbok look quite a bit like our pronghorn antelope, the blazing speed of which was molded by a North American cheetah long since extinct, but with our pronghorn still possessing that superlative and now useless excess of speed, like an echo, a ghost or shadow of what once was.

I want to see what these babies can do. Dennis has pulled over to the side of the road to take photographs, and once he finishes, I climb out of the truck and walk across the gravel road and into the basalt, where the springbok stand in a loose shadow, like a school of fish, scattered yet coalesced, watching.

A few twitch their tails and look about as if for guidance or clarity while I continue to approach, but for the most part, the general mood is that they do not take me seriously.

I'm having to walk carefully among all the rocks, and though I suppose I must appear to them exactly as I am, and that they are perceiving reality—a forty-six-year-old man in tennis shoes, walking clumsily through the rocks—in my mind, I am twenty-six again, or even sixteen, and I am remembering the fastest I ever ran, and the strongest I ever was, and it is a pleasant memory, a pleasant myth.

I continue to wander closer, and only when the nearest animal begins to take its first step away—more of a mechanical reaction, like the end of a magnet rotating away from a polar similarity, than any true fear response—do I lurch into a sprint, charging straight out at the herd.

In my mind, I'm flying; in my mind, I'm fleet, and to be feared, so that this sudden announcement of my desire should be enough to catapult them into wild and springing flight—eliciting in their terror an involuntary burst of full throttle.

Instead, for a long time, they do not run—do not move at all—but instead continue to watch me a little anxiously, though with what might be viewed as concern. As I jar my way toward them, navigating the rock field as athletically as I can (in my mind, I'm still flying), it appears for a moment in that dimmer, redder light that it is *me* they are concerned for, rather than themselves.

And though it gradually begins to dawn on me that I am not moving quite as fast as I had envisioned—again, that distance almost a chasm between truth and fact, or perception and reality, the living, waking dream-space between—I *am* gradually drawing closer to the herd—quite close, actually—so that finally they are compelled to begin to move.

They do so, however, not with the electric and involuntary surge of electricity I had envisioned, but with all the excitement and motivation of a herd of old donkeys, leaning forward into a trot; and still I stay after them, dogging them, but now they are spreading apart, immediately and effectively diffusing my charge, I see, so that now only one or two of them will have to sprint, if that.

I used to be fast! I used to be a lion. Wasn't this how it once was, or am I imagining that? Was it but a dream? I veer suddenly and angle in on one of the nearest springbok, closer now than I have yet been, and it seems to be my erraticism rather than any blaze of speed that concerns the springbok, for it wheels away with a couple of quick steps—for a second, I get a glimpse of the pre-speed, or warm-up speed, of which they're capable, if not the real thing. And then, almost, it seems, to humor me—as if understanding what I want—the springbok leaps into the air and kicks its spry little legs, *boink, boink,* almost as if unable to help itself, like a sneeze, or a shudder.

Even with that mild response, the springbok has doubled its distance from me with no more effort than a yawn, and the rest of the herd has also ambled up into the rocks, a similar distance away. I stop and watch the springbok, which have also stopped and are likewise watching me, and I have two choices—to be amused or frustrated by this seemingly uncloseable distance.

I decide to choose the former and to remain in the comfortable territory of perception, if forced to lean one direction or the other. And the springbok know this, knew it even before I got out of the truck, I suppose. A man on foot is neither a lion nor a cheetah, and desire is not substance; desire has neither fang nor claw—is nothing but a kind of ghost.

I walk back to the truck (the springbok continue to stand there on the hill, watching—logging this curious behavior into their databanks, perhaps) and pick my way carefully through all the rocks, grateful that I did not trip and fall and injure myself. Such an injury would have been bad enough, but explaining it would have been even worse.

We drive up and out of the quick-darkening springbok canyon —it looks like the Hollywood setting for every stagecoach ambush that was ever filmed—and follow the winding road up onto a mesa. I had thought the fast-dimming sky was merely a function of our being down in the canyon, with the high basalt ridge simply blocking out what was surely still a surfeit of remaining sunlight, but upon gaining the mesa, I am surprised and disoriented to find that even up high, a similar darkness pervades, that although the heat of the day has been summer-like, or beyond summer—like some new-created fifth season, incandescent and broiling—there is no such thing as twilight, and it's possible here to see night falling like a bird shot from the sky.

We stop on the mesa and stare down into the next broad red valley. There seems to be no soil all the way to the horizon, only red rock and gravel, in which grow occasional tufts of straw-colored grasses, or, in certain clefts where a trickle if not a torrent of rainwater might occasionally flow, a tree or two, dramatic and lush in their solitude, and with that clump of feathery green soothing against the baked red world that spread in all directions.

There is a hut on the mesa, assembled of flagstone and generous scoops of mortar—more of a small pavilion than a hut, with a single crude-hammered table onto which have been carved and then painted the markings for some chess- or checkers-looking game unfamiliar to me. Countless bottle tops are heel- and time-smashed into the red grit hardpan in all directions—every available bottle-cap-size space on the ground is embedded with such a cap, the printing on each long ago worn down to bright spit-shine polish, so that were a light rain falling, the whole embedded hillside might be as slippery as lake ice, and those who sought to walk on it might be able to glide with the smoothness of skaters.

But this windy evening, with inky night settling quickly over the rift-valley beyond, the mesa is bone dry, and Dennis and I sit in the abandoned open-air shelter and watch the darkness come hurrying in. The day's last light glints upon the bottle caps, giving the ground the brief appearance of a glittering diamond dance floor, and then that dazzle fades and it is darker than ever, and an evening wind begins to escalate out on the point.

This, as well as the view—the summit, the common juncture between several valleys—must be why this point was chosen for the hut, and Dennis and I take a cold Tusker beer from the miraculous little refrigerator and sit there in that last light, with the heat finally leaving the land—leaving it quickly now, in that wind, disappearing almost as quickly as the light itself—and we stare out at the valley below.

There is nothing, and we look out at nothing, and we feel a peace and a stillness that in no way should be correlative to the near past, the human history that resides just beneath the land to the north, the land we are looking out at, only a few short decades ago; or, for that matter, yesterday, or this morning, this afternoon, now.

For a long time, beneath the general veneer of my pop culture, I'd heard occasional abstract allusions, like whispers, to Africa's having endured a genocide no less horrible than the Holocaust —the erasure of a race of mankind upon his and her own homeland for no other reason than . . . well, for no reason; for artificial external factors masquerading as reasons, flimsy masks meant to cover, like manhole covers, the horrific vents leading up and out of the twisted, rotten places within the human heart, human mind, human soul. Vents and rifts arising as if straight from some hellish cauldron far below.

Here too, the body count—as if one million was somehow worse than one—was several million, and also occurred in the twentieth century, often at the hands of the same German military. There are too many obscure textbooks and oral history compilations detailing the atrocities, and yet there are not enough.

There are an infinitude (if one cares to look deeply enough—which is, again, not very deep at all) of heart-rending, or perhaps heart-deadening, descriptions of the barbarities committed by one human against another, one human to another. One of the favorites of the day—like a trend, perhaps—involved tossing a child, or a baby, from one soldier to the next, until the soldiers' arms grew weary, or they were bored, at which point one of them would end the fun with their bayonet.

Not once, or twice, but thousands, then tens of thousands of times, then millions, on each continent, like some mad cell replicating. It was this way in Europe; it was this way in Africa. It was this way on the American continent as well—we were just fortunate, perhaps, for our peace of mind, to have gotten much of our killing and torturing out of the way in the previous century. Or so one must hope. An extra half-inch of steel upon the manhole cover; another few years' remove. Or so one must hope.

In his startling and lyrical memoir of East Africa, *The Zanzibar Chest*, Aidan Hartley recalls growing up among villages in which, as recounted by his father, "the last Caucasian they had seen before him was an officer of the Schütztruppen who had ordered the village chief to be buried alive simply as a warning to others to behave."

Amid the veneer of the times, there is often heard, whenever the uncomfortable subject of the still ongoing genocide in Africa is broached, the mention—not a rationale, certainly, but more a pitiable kind of defensive attempt at even another half-inch of detachment—that the current African wars, like many of those in the past, are civil wars, and therefore somewhat of a lower priority on the long list of "trouble spots" to which a white knight on a steed might gallop, seeking to spread freedom.

A central façade to Europe's colonization of Africa was the

notion that Africans were "incapable, childlike, vicious, and primitive," explains Hartley. It was an old and self-fulfilling truth. "Numbers had already been weakened by the nineteenth century Arab slave trade," writes Hartley.

> With Europeans also came sand flies . . . which infected Africa's soils with jigger worms that rotten the feet of the barefoot peasants. Colonial forces . . . imported cattle infected with rinderpest. The disease spread in a wave from the Horn to Southern Africa, destroying multitudes of cloven-hoofed animals in its path. Smallpox, syphilis, and a battery of plagues from the outside world followed. The Africans who survived, decimated by famine, went to war over what resources remained. East Africa is dotted with monuments to the conflicts and pestilence of that time, such as the Rift Valley town of Eldoret, which means "The Place of Killing." As the Europeans ventured farther into the interior they discovered swathes of territory where few people survived. My ancestors beheld this scene and assumed that it had always been like this—with the Africans living in a benighted state of perpetual war, pestilence, and famine.

Against such a tableau, wrote one of the chief European proponents of colonization, Captain Ewart Grogan, it was the duty of so-called civilization "to fit the country as a future home for [our] surplus populations . . . to make new markets and open up country . . . to suffer temporary loss for the future benefit of humanity."

Some of that temporary loss involved the practice of desert-dumping. The colonizers soon learned how tedious it was to execute entire villages, and problematic as well, for health and labor reasons, to dispose of so many bodies in the heat. They began to export vast numbers of their prisoners, by truck and by train, into the interior of the harshest and most inhospitable land in

Namibia—the territory once known as Damaraland—the lunar desert landscape that has been unchanged for 182 million years, with its nearly nonexistent moisture and otherworldly temperatures. It was a place where no humans, or no culture, had ever been able to survive, and it was the haunt, the birthplace, of the desert or black rhino.

The Germans dumped the prisoners there, in rags or nothing, with neither food nor water, abandoning them to die, which they did.

Some, however—miraculously, torturously—lived, and remained there, hanging on, like seed drift having made it all the way to the moon, or Mars. Their descendants—second- and third-generation survivors now—still dot the perimeter of Damaraland, in knots and clusters, herding goats and standing ankle-deep, and then knee-deep, in the wave of AIDS that is submerging the continent, latest and probably not last of the legacies of mankind.

Travel far enough back in time and almost anywhere in the world was once a paradise—a greening Garden of Eden—the sagebrush prairies once submerged beneath shallow swamps and marshes, the austere mountaintop once a riverside meadow. But Damaraland is not one of those places, or not within the last 130 million years, anyway. If anything, it is the anti-Eden, and yet somehow—as if, somewhere in the world, there existed such an overflow of grace that some wisp or trace of humanity would be allowed to survive here—some of those deportees found a way to keep living, if not to prosper.

Perhaps the nearest water hole was five days and nights away, its whereabouts and existence unknown to them. Perhaps some of the travelers gave up, and holed up, there in the hardpan sun with no shade or sustenance, and after two or three days simply

died. Others might have set out walking, in one direction or another—west and north, perhaps, so as to keep the unbearable
sun at their back—where after a couple of days they too expired,
baked like hams or roasts, their internal organs cooked to the
point of failure.

Against all known rules in the world, however, some of them
survived. They held on long enough, one way or another, found
enough moisture or made do without, until becoming fortunate
enough—lucky enough, guided almost always by grace—to
stumble blindly and unknowingly toward, and then into, one of
the hundred or so alkaline watering holes that dot, like tiny water fountains in a school hallway, a territory of more than 50,000
square miles.

The traveler, or travelers—a band, a cluster, fast diminishing,
their numbers falling with each passing hour like individual petals, drying and curling to a leathery crispness, fluttering from an
aging blossom—would have wandered across the trackless, stony
plains. Only from the highest promontories would they have
been able to discern any larger pattern in the landscape: more
mesas, more plains, more absence, more suffering.

Some mornings, to the southwest, they might have noticed
lingering on the horizon, well over a hundred miles away, a thin
silver ribbon of opaqueness, something like a long wall of cloud
that might have hinted at rain but that never approached and indeed vanished by noon every day. That would have been each
morning's fog back out over the Skeleton Coast (so named for the
countless shipwrecks), unreachable by foot, too distant, on foot,
and with—unimaginably!—even more inhospitable country beyond the rocks and heat and aridity of Damaraland, in the form
of the great wind-whipped sand dunes, the largest in the world,

that separate the coast from Damaraland like a skirt or apron of superheated and all but lifeless sand. A gecko here and there, or a sand snake, and mountains of towering, shifting red sand.

Those who saw and were lured toward the shining silver and temporal cloud, elusive, ungraspable as a dream, surely perished, while those who stayed and made their peace with the desert might have had a chance, and *did* have a chance—the chance of grace.

Staggering, tripping and falling, they would have moved slowly across the stony red landscape. In the distance, they would have seen the shining silver backs of the rhinos, immense, at even a distance of several miles and against a backdrop in which there was nothing else to give them perspective. The rhinos, their broad backs and heads and horns reflecting the sun like mirrors, would have appeared to be floating across the landscape, drifting not unlike great clouds themselves.

Other times, the rhinos would have been much closer. The rhinos are extraordinarily nearsighted (for who, or what, needed vision in a landscape in which nothing ever changed, and in which there were so few challengers—so few of anything?). In a land of such heat and grit and wind, might the resources required for the maintenance of ocular precision prove to be extravagant?

Almost always, it seems for the rhinos—for everything—there were fierce choices, fierce pathways to survival: always the gambler's choice, and yet always the right choice, or the lucky choice, or both.

Keen vision or no vision? Toes or no toes? One horn or two or none at all?

As if backtracking only a short distance, it seems here that the puzzle pieces of time, as well as of consequence, might yet be reassembled to bring reason and meaning, with such order reappearing almost as simply as geologists reassemble the fractured plates of old Pangaea itself across time.

There, the abandoned travelers—one surviving out of every hundred, or maybe two out of every hundred—an Adam and Eve—might have laid their fevered bodies in the shallow waters, the tiny oasis stippled not just with the cloven hoofprints of prey, but the dinner-plate-size tracks of the rhinos, the bicycle-tire-size imprints of elephants, and the fresh mud-spread long-toed tracks of hyenas, wild dogs, and jackals; the firm, muscular tracks of cheetahs and leopards, and, most terrifying of all—terrifying even to a man or woman dying of thirst—the sprawl of lion tracks, fresh-made and larger than the largest man's outspread hand.

Somehow, the travelers, the prisoners, survived. The lion had passed by an hour or two earlier and did not return until after the travelers had retired to a cave, or fetched up to some point higher on a hill, where perhaps they made a small cairn in which to huddle. A spear was fashioned out of a green branch taken from the oasis, and sharpened by rubbing it against stones. They went back for water, they scavenged, they killed, and in turn, some of them were killed, and yet somehow enough of them kept living —chosen, one could say, to remain, like an uninvited but stubborn guest, on the landscape: scratching and clawing and suffering and enduring.

Who can say at what point residency becomes securely established, if ever? If the guest did not arise as if from an on-site excavation, like the rhinos—constructed of the very dust and clay and mud and sticks and straw upon which it stands—can not all species, whether great or small, be thought of as a kind of guest, with their biological authority in that place a measure instead, or alternatively, of the relative grace (or absence of grace) with which they fit and inhabit that landscape?

On either count, of course, the rhino, and so many other of the African species, possesses the fullest possible and most authentic—if no longer durable—citizenship in the world. The rhino—born of this soil, and of the fire beneath these rocks—is fitted to the surface above as magisterially—as magically—as is our own species almost hopelessly ill fitted, clumsy and almost completely unauthorized.

This landscape is too severe, too stark and beautiful, for the observer to believe in luck. Here, there can only be infinite design, or grace, or some beguiling and elusive and non-knowable combination of grace and a greater desire.

What must that gambler's choice for survival look like, each time it presents itself? Does it appear as an opportunity, or an ever-narrowing absence of opportunity, a bottleneck through which there is now almost no chance *but* to hurry through, shooting the gap before the chance closes, taking it as if diving through an open window?

And once the decision is made—deciduous leaves and seed production, or coniferous? emerald breast plumage, or sky blue?—and a species passes through that bottleneck, or crests and then crosses the summit of that tipping point, it must appear in retrospect as, again, a forced move in a designed space.

So seamless is the new fit (the rough edges pasted over by the smoothing polish of time) that there seems never to have been a gamble, there seems never to have been any critical tipping point, in which one step meant oblivion and another salvation.

Who or what makes those choices, and where is the hidden roster of failure, or wrong choices? If the Earth itself is the Book of Life, where is the tablature of loss?

With the exception of deeper Damaraland, the region toward which we are traveling, the landscapes are not overwhelmingly different, and neither are the climates—not by quantum orders of magnitude, at any rate—and yet the choices, the differences, are huge. In northern India, in the Himalayas, I had been astounded to find pueblo-dwelling cultures in landscapes at high desert elevations that seemed the precise equivalent of the much lower elevation deserts of the pueblo-dwelling cultures of the American Southwest, with even many of the religious practices and dances almost identical, shaped not so much by time, it seemed, as by the land. Convergent evolution, scientists call it: different species adapting in similar ways to fit similar niches on earth.

And for a long time, that was how I saw the world: as if those points of difference, those opportunities to diverge from the design or momentum of the world, were infinitesimal at best. Almost as if the dice were long ago cast and the outcome all but preordained.

A single day in Damaraland can explode all that. Certain landscapes appear to be no more than a degree or two off-center, when in reality, here in Namibia, the variations in the paths of life might possess differences of ten, fifteen, even twenty degrees, so dramatic are the outcomes, and the cascading subsequent effects on the web of life.

Something as simple as a leopard caching its kill high in the breezy limbs of an acacia tree is *huge;* it allows the leopard (sundappled-leaf-spotted) to avoid unnecessary conflict with one of the other great and even more muscular cats, the African lion. How paradoxical it seems that in such a harsh environment, life —which we are led to believe is rare and precious—should be so ever present and excessively tenacious as to occupy vertical niches, towers and strata of life constructed seemingly on thin air. Leopards floating in another world ten and twenty feet above our heads.

One of the keys, the secrets of such fantastic life, is the energy of the sun, the excess of the sun—the fantastic solar radiation in Africa yielding a nation of grasses, on which so many ungulates can feed, on which then so many fantastic meat eaters can dwell.

Enter the rhino, then, like Exhibit B, or perhaps Exhibit Z, born magnificently of the loneliness of Damaraland—as if this was one great thing that landscape decided to make, concentrating its wonder and possibility and biomass not in an overly crowded scrabble of multispecies layering, but instead in the one great work, *rhino*.

Where are our rhinos, I want to know—America's—and where, for that matter, are Africa's bears? Could not a grizzly bear live along the cold and foggy Skeleton Coast, foraging on crabs and fish and whales and seals? And could not a rhino graze wherever a bison went, and beyond, climbing up into the rocky basalt plateaus around the Snake River canyon, with a North American sun glinting brightly on its leathery hide?

This is one of the beauties of Namibia, and rejuvenating to a middle-aged traveler: the reminder or revelation of an entirely other world, even as the traveler continues to glimpse similar stories, here and there, decorated in different cloaks, different stripes and spots.

Are time and landscape conjoined—is there a visible connection between time and space—or are there sometimes gaps or differences, certain places in which one force, such as time, exerts a greater guiding breath upon the world, while in other places the made and physical world of space, and of landscape, possesses the greater authority?

Are there places too—such as Damaraland—where time and its passage plays almost no factor at all: where the world's essence and mystery is all about the land, and only the land, forever unchanging?

Across such wretched stone prairie, completely bereft of hope —for if there is no passage or movement of time, how can there be hope?—those abandoned prisoners would have crept and crawled and staggered, following the tracks of the solitary rhinos, or even, sometimes, walking near and alongside them, as if accompanying them, staying just out of their vision, the stumbling prisoners not born or unearthed or excavated from this land, but having arrived as clumsily as seed drift.

Surely the travelers might have thought such an immense animal needed water, lots of water, and though the distant sight of the horned behemoths might have seemed a dream, the fresh-scuffed tracks on the red dust and gravel were not a dream: evidence of where a thing, an amazing thing, had passed. The travelers might have noted the frayed and frazzled ends of the giant *Euphorbia* bushes, dome-shaped bushes, sometimes eight feet high and twice as wide, the branches still glistening slightly from the rough chewing of some animal: though if the travelers themselves tried to gnaw on the ends of the branches, they would die even more quickly than if they had not, for the *Euphorbia* bushes are toxic to lesser mammals.

How the rhinos themselves survived is a matter of conjecture. Where once they spread across the sub-Sahara portion of Africa in great numbers (some estimates place the peak of their population around 100,000—the same estimate, coincidentally, for the United States' once-upon-a-time population of grizzly bears), the rhinos plummeted in the twentieth century of war.

They had no substantial natural enemies, up until the twentieth century's distribution of firearms. Rhino calves were to be defended against packs of hyenas, but that was about it, until the rifles came. Their horns had been prized for use as ceremonial dagger handles, particularly in Yemen, and had been used for centuries in China as a medicine, in ground powder form, for various illnesses, including the flu. (The alleged use of rhino horn as an aphrodisiac appears to be curiously a Western fabrication.) But until the advent of modern artillery, the rhinos' numbers were relatively unthreatened. The horns they carried before them (actually composed of densely matted fur) were an experiment that had allowed them to survive in the world for so long—a far more effective weapon, it would seem, in that regard, with fifty million years under their belt, while we with our repeating rifles and plastic bullets and neutron bombs and tasers and lasers and rockets and machine guns, have survived for less than two hundred thousand.

The horns were vital to the rhinos, valuable if not priceless to their survival out in the desert; it was just plain bad luck for the rhinos, rather than an ecological miscalculation, that we would appear from out of nowhere—fifty millions years late to the party—and decide, for no real reason at all, that we desired the crafted thing.

Considering such things, you can't help but wonder, why are we here, and why did we get here so late?

Some teachings would suggest love, or even loneliness, is our calling card, the thing that got us to this banquet—that a creator wanted our curious company—and indeed, that might be the only way ultimately to explain our membership in the world, and our sudden and frail and hairless propulsion through our own bottleneck of the elements—snowstorms and desert heat, wildfires and glaciers, lions and tigers and bears, viruses and plagues—

into the pantheon in which we find ourselves today, billions of us adrift and estranged.

The sixth day, indeed! And watching us paratroop down to earth, spinning perhaps like dandelion fluff, what other final caveat could the creator have uttered to us: the do-no-harm entreaty, or one of stewardship?

Dominion, the creator whispered, according to one telling. But that seems a long time ago, and because the command, or so we tell ourselves, was not shouted, we choose, I think, to view it as a blessing rather than a warning, and a right or a privilege rather than a responsibility.

Perhaps—according to at least one telling—we are, in the time-honored tradition of love, beginning to take advantage of that love.

It must have seemed that way to the rhinos, that their birthright had left them. They began to be felled in great numbers, with a ton or more shaking the world every time they fell. They were poached by soldiers of the last century's various civil wars that shredded southern Africa—poached, it is said, by both armies, all armies, in the conflicts. They were killed by tanks, by machine guns mounted to jeeps that chased them as the rhinos galloped at thirty and forty miles an hour across the desert beneath the broiling sun, with their horns—the larger front horn generally weighing between seven and twelve pounds, and the smaller, second horn, perhaps half that—cut away, after death, while the other three thousand pounds of majesty was left rotting or baking in the sun, testament to a new and mindless, surely uncrafted tipping point.

Across southern Africa, the armies trafficked in illegal diamonds, ivory, and rhino horn to help fund the great Cold War battle between communism and capitalism that was overlaid on the existing civil wars brought by resource depletion and the injustice of apartheid. Communist-backed Angola, to the north, having war waged against it by the U.S.-backed South African army to the south, used the middle territory of Namibia as its battlefield. Black rhino numbers swooned, falling from several hundred thousand before the twentieth century to below 2,500 in the 1990s: one of the ultimate bottlenecks in the history of any species great or small.

Images of Noah's Ark return: a population whittled down to minuscule populations. I think once more of the grizzlies in the Yaak—a dozen?—and I want desperately to believe that the rhinos can cross over this latest tipping point and continue on into the bright haze of the future.

Apartheid was finally broken, and the wars waned; the rhinos were able, there at the bitter edge of extinction, to take a breather. They were still being poached—the South African army had issued thousands of .303 rifles to the nomadic goat herders living at the edges of Damaraland, ostensibly for protection against any Angolan invasion—and there was a greater and longer-running war of resource depletion, and the ecological ruin of desertification—livestock, and humans, dwelling overlong in any place of vegetation, leaving behind nothing but sand and wind-scalloped erosion, as ruinous as any bomb crater—but at least the rhinos were not being pursued now by the armies.

Pushed to the brink, the world conservation community began to do a little imagining, a little crafting and designing of its own, dabbling in the power of paradox.

They hired poachers to look after the rhinos. They made the rhinos more valuable to the poachers alive than dead. There were so few rhinos that they could assign one poacher—one game guard, as they were now called—to each rhino.

The game guards followed their charges through the desert, tracking them and protecting them, keeping them almost always in or near their sight, like guardian angels with heavy AK-47s slung over their backs, and every bit as attentive to each rhino's movements, now, and those of other humans coming and going on the landscape, as they ever were. More so, now, for the rhinos' presence brought sustenance.

Not all of the game guards were poachers or ex-poachers. But that was who knew the rhinos best, and who could track them the best: amazing trackers, able to follow the rhinos across bare stone. If it is true that in the seeds of everything's creation lies also the story of its destruction, then might it be true too that in the seeds of a thing's destruction lies similarly its salvation?

A second thing the world conservation community did, in their desperation, was to try to get to the rhinos before the poachers did, and to gather them up for safekeeping, placing them in the gigantic Etosha National Park, a 8,600-square-mile landscape entirely surrounded by a chainlink fence. It's Dennis's and my goal to catch a glimpse of this park near the end of the trip. Although there is poaching even within the national park, it is less pronounced, and somewhat easier to defend. I found the notion of fences distasteful, even as I recognized the rank luxury, back in Montana, of being able to indulge in such an attitude. My valley still has its twelve or so grizzlies, with the poaching—the illegal killings, for killing's sake—continuing there seemingly unabated. My prescription, however, is for greater protection for the vast, or once vast, expanses of wilderness in which the grizzlies are still secure. In a forested landscape—unlike the saltpan playas of Eto-sha—the bears could find hiding cover in any direction, rather than being shiningly visible—luminous—at distances of up to ten miles, as the rhinos are on the salt flats.

How would I feel were that my absolute and only rock-bottom chance at preserving the species—to put them behind the safety of a chainlink fence? I know desperation, even now, with regard to the Cabinet-Yaak grizzlies, but how would I feel to know the calm chill beyond desperation: the knowledge that this alternative—so nearly an admission of failure, and a capitulation, in itself—was the only remaining choice, other than extinction? To build a *fence*, as if enclosing a garden rather than a wilderness?

Would that not just as well be a thing so much like extinction —so much like failure—as to already render the species gone? For no species exists isolate and separate in the world, not even our own, but is instead always to some degree a function of its relationship with the land that supports it, and when that land and that relationship is gone, then so too vanishes the heart, essence, and true identity of the species.

You can round up the last of the last and place them in corrals and perform captive breeding programs. You can extract DNA from their tissue and store it in vials on the shelves of freezers for the day when scientists attempt to restore and re-create life with no more aplomb than that with which they might embark on a shopping expedition: browsing once more, as if in a mere catalogue, through that Book of Life.

How would I feel were that the only choice left to me and to my landscape? Would I finally acquiesce and say, *Go ahead, round them all up*? Or would I try to ride things down to the bitter end, never giving up, always holding out some last hope? "Lord, let me die . . . but not die out," writes James Dickey in his poem "For the Last Wolverine."

If absence-from-landscape is its own kind of extinction, then I believe that would be my choice—to try to hold on at all costs, even if diminished—for there would always be time—an eternity—to begin the repopulation or recolonization efforts from captive stock, or zoo stock.

It would not, however, be an easy choice. It is not an easy choice.

And if a rhino without its landscape is not a rhino, and a landscape without its rhinos renders the rhino likewise functionally extinct, what is a rhino without its horns?

That too was one of the desperate proposals and, for all we know (we may never know), one of the partial solutions: for biologists to hunt down the last black rhinos and, rather than trying to herd them all up and transport them to Etosha—a budget-busting and logistically problematic effort—to instead simply beat the poachers to the treasure, and to saw the horns off themselves, placing the horns in secret safekeeping, government vaults, where they would be kept off the world market until such possible points in the future as the rhinos were recovered to sufficient numbers for their horns (like the ivory tusks of elephants) to be legally traded.

Chopping off their horns—the very thing, the one thing, perhaps, that most identifies them as rhinos. What if the choice were to chop off their legs, then? Why not chop all of them into various parts and pieces—hiding the parts then like the pieces of some rudely deconstructed puzzle—freezing them, perhaps, and then awaiting an eternity for them to reassemble?

What kind of love for a species—or for anything—would it

take to make such a decision: to separate a thing from its home, and from the very essence that had made it?

In *Horn of Darkness: Rhinos on the Edge,* Carol Cunningham and Joel Berger describe their time spent in Namibia in the mid-1990s, trying to research the effects of the horn removal program. The tale they tell is of entering naively a matrix of various environmental groups and philosophies, both foreign and African, replete with the egos and selfishness that seem to attend almost any of man's pursuits. Berger and Cunningham weren't convinced one way or the other that the horn removal program was working, and tried to craft a scientific study that would measure one aspect of horn removal, which was the possible reduced ability of female rhinos to defend their young against hyenas.

The need for a larger study — on whether horn removal aided the entire population — was made clear to them, for there were so many variables: wars stopping and starting, horns regrowing, not enough funding to study the animals, and an insufficient number of animals to make a statistical study. It seemed that anecdotally, for every story of finding jeep tracks showing where poachers had followed a rhino and then turned away, seeing that it had no horns, there were stories showing the opposite, where a rhino had been killed anyway: perhaps out of spite, or perhaps for the nub of horn that had started to grow back, with the keratinous material reforming at a rate of about six centimeters per year in the front horn and three in the back.

Again, at what point do we walk away and let the rhino live or die, if we refuse to do the hard work of allowing it to retain some big wild country in which it can be safe? Maybe not closing off a swath of country, making it off-limits to people entirely; but designating some, certainly, as places where motors cannot travel, and where if a traveler wishes to cross the desert, he or she must do so on foot, or mule, or camel, or some other fashion that does not jeopardize the time-honored wildness of the place and all of that place's complex inner workings, both known and unknown. *Wilderness.*

It comes back I think to that question of love, and love's nemesis, denial. If we insist on forever following after the rhinos, clipping their horns monthly as if pruning a hedge, are they still rhinos, or have we already allowed them to vanish and are simply not admitting that to ourselves?

The horns won't be gone forever, the horn-snipper will answer, *just long enough to get them back on their feet, to shoot that gap, to recover. Just long enough to buy time. Just long enough for one more chance.*

To one who loves rhinos, and who can summon a hope for the future, and who is too keenly aware of the difficulty of establishing a big wilderness—a huge wilderness—much less enforcing it, it is a compelling, if not heartbreaking, argument.

What would I do back home? Would I remove the grizzly's claws to save it, or to *maybe* save it? Would I erase the bull elk's antlers in autumn, or remove the wolf's larynx so that it could no longer howl?

And again, considered against the arc of the near-coming future, are not all such deliberations quaint and obsolete? With designer genes, and all the coming experiments—the insertions, splicings, and takings-away—so imminent, perhaps a lover of wild country—and indeed, a lover of the earth—might ought to look hard one last time at his or her beloved, and at the amazing puzzle that once was assembled: for are not the child-gods even at this moment disassembling the very foundations and underpinnings of that once made and once dreamed world?

In the end, it might have been a combination of factors that conspired to begin easing those surviving rhinos—not unlike the few surviving prisoners of war who were dumped in Damaraland —through that constriction of death and back into the expansion of life.

Or is it the other way around? Is it the brief, concentrated moment of life that is the constriction—the beautiful bottlenecking, beautiful confluence, of extraordinary moments, elements, and circumstances—while it is death, extinction, and absence that is the boundless broadening, the falling-away of all borders?

The rhinos were here, then gone. And now—maybe—they are coming back.

Namibia was here, intact, then war-torn, stranded between civil wars, but has now gained its independence from white colonial rule—the last country in Africa to achieve independence, which it did in 1990, removing itself from South African rule —and now the rhinos, maybe, are coming back: as if arising yet again from some place of burial, some place of temporary safekeeping.

Maybe it was the end of the drought that has helped them recover, along with the cessation of war, along with the fierce patrollings of anti-poaching units (in Zimbabwe, more than 150 rhino poachers were killed in the act). Maybe it was these things, combined with the horn removal program, as well as the worldwide ban on the sale of rhino horn. (Though there are those who argue, fiercely, that the latter conspires to create poachers, empowering them to rule their lives in pursuit of the fruit, the horn, which, if legally grown on farms, and sold and traded, might see its worth decline to a point at which poaching might no longer be the career opportunity it once was.)

And waiting in the wings, watching all of this, are the wealthy so-called sport hunters, the fellows who have killed almost everything that leaps, crawls, slithers, or hops on the African continent: slayers of zebras, giraffes, hippos, elephants, and, for all one knows, even the mighty ostrich, with their sights now set lasciviously on the rhino. It makes me sick and angry to consider it — it frightens me to consider the seed that might exist within a fevered brain to desire and labor toward such a goal — and it angers me also that the rhino conservation community is so incredibly impoverished that the year-by-year survival of the Namibian program might be made or broken by the sale of a single rhino killing permit, for the amount of a paltry few tens of thousands of dollars.

Not for the hunting permit's fifty or sixty thousand dollars could any god among us design, engineer, and then manufacture such an amazing thing as a rhino.

I'm a hunter myself and am familiar with the criticisms, and the seeming paradox, of the act. But to want to walk up on a nearly blind creature, something as otherworldly and prehistoric, something as made and fitted and beautiful as a rhino, and to dispatch it with a high-powered bazooka masquerading as a rifle — such a desire seems to speak, from what I know of human nature, to either a twisted heart of frightening blackness or, more generously, to the undeveloped, often thoughtless governance of the mind of a boy.

And if the latter, then to my way of thinking, there are other, more constructive ways to be a boy.

• • •

It's long been the Holy Grail of conservation strategy to devise plans wherein the preservation of a species becomes a dominant and obvious economic value. So *community-based conservation* is the current buzzword, the cutting-edge activism that philanthropists are exploring and pursuing, these days—searching for pilot projects, for experiments that combine a meld.

A little group with a little budget, the Save the Rhino Trust (SRT), founded by Blythe Loutit and her husband, Rudy, has been working on this hard rocky ground for twenty years now —an astonishing amount of time for so small a group—in what, although it would be nice to avoid a war analogy, can only be called the front lines of rhino conservation. They've watched the rhinos spiral down toward the oblivion of nothingness and have helped pass the rhinos through that bottleneck, and are over on the other side, in what will hopefully one day be called recovery.

Their field station in Palmwag is where Dennis and I are. We've left our little rest area—still not fully acclimated to Namibian time: to our bodies, day is night and night, day—and we're journeying, traveling now beneath the desert stars, different stars, or rather, a different sky. After hours of driving without seeing another vehicle, we come to a chainlink gate stretched across the dusty road, with a little hut perched next to it. This is the infamous rinderpest fence, erected decades ago and stretching across all of Namibia, a seemingly arbitrary and invisible bisect below which livestock, and in theory, wildlife, is quarantined and may not be exported, while livestock (and wildlife) north of this flimsy fence is deemed healthy and acceptable.

There's no one in the little hut, so we get out and open the gate and drive through. We stop then, close the gate behind us, and drive on, suddenly on soil that, in theory—as if through divine revelation alone—has been declared to be a rinderpest-free zone.

Some miles down—a couple of wrong turns here and there, in which the road turns to a trail and then ends in stony desert, nothing but stars and stones—we find the road Dennis is looking for, marked by a plywood sign with the neatly hand-painted silhouette of a rhino on it, as declarative and yet ambiguous as a pictograph.

We turn down the gravel road—a trail, really—and eventually encounter some thatched-roof structures.

Almost everyone is down at the campfire, and we join them: Mike Hearn, field director for SRT at Palmwag, and Dennis's Round River students, who are doing conservation biology field studies in conjunction with SRT, and the Round River instructor, Jeff Muntifering. It's both awkward and welcoming-feeling to be coming into a camp of strangers late at night—affiliates, if not quite kinsmen, amid so much other space and emptiness, and so far into the evening—and we squint in the firelight, trying to better make out the faces of the students, who are introducing themselves; and here too, again, it feels as if everything is new and yet everything is ancient.

Dennis asks the students how things are going, what it's like to be out here, and what they hope to get from their studies. I try to focus on their goals—*preserving endangered species, empowering local communities with conservation abilities and incentives,* and so on—but what stands out for me, I have to say, is the story about all the zebra snakes—spitting cobras named for their black and white markings—that keep showing up in camp: three thus far. Just last week, one student says, she was in the shower stall one night—an open-air, sun-warmed-water gravity drain system, with pea gravel underfoot to absorb the water—when she felt something wet strike the back of her leg. She turned and saw that a zebra snake had spit at her. She covered her eyes, which are the preferred and lethal target for the snake, and was able to escape. Though I like to think that I am not overly frightened by a wilder nature, I'm surprised by the chill on my spine.

And not long after she has finished her tale—told with the nonchalance of youth—than another of the students discovers a scorpion in our midst, scuttling between our various besandaled feet. The scorpion is plump, the color of a dark plum or raisin, or a bruise—I imagine that in the dark, color is a function of the poison within—and the students scoop the scorpion up in a tin can, examine it briefly, and then carry it back out into the desert to release it.

So far there is nothing about this place that makes me feel I fit here, and yet there is simultaneously a fast-deepening allure, as well as a hunger, that is surprising in its power. Gradually, I come to understand that everywhere I look—any place I turn my head—I will see something entirely new to me, and perhaps inexplicable.

Gradually—and then more quickly—I have the feeling of awakening, all the more strange for not having been previously aware of the sensation of sleeping.

Mike Hearn—tall, movie-star handsome, with his long brown hair sun-bleached, dark brown eyes, friendly smile—sits quietly and listens, letting the students talk. When he does speak, it's in a bantering way, with the accent of a British diplomat, even though he's been out in the desert now for eleven years, having come here straight out of college, searching for the source of rhinos when he was barely twenty-one.

The Save the Rhino Trust folks didn't believe he'd last—just one of any number of drifting young men searching for a cause, whose interests and commitment would surely shift and wane over time, and they shuddered to think of the liability, as well: placing a city boy out in the middle of the Namib Desert, unarmed, to study and defend the rhinos.

Instead, they agreed to keep him in the office in Windhoek: a place that had begun to unnerve me after considerably less than twenty-four hours. They kept him there for a year, in their offices, filing papers as a volunteer, and still he would not go away, still he was determined to live among, and help, rhinos.

So at the end of that year, they released him back into the real world. He set up and executed field studies that determined more precisely the diets as well as feeding habits of the Namibian black rhinos, and he helped establish protocol for the visual identification of the handful of rhinos that remained. He wrote grants and hosted passers-through from the scientific community, and lobbied his old government to take an ever more active role in supporting the resource, for the resource's sake, and for Africa's sake, in a land Britain had once squandered, had once owned or perceived they had owned during the British-German colonial days (which had followed hard on the heels of the United States' slavery days).

What are any of us doing here? How dare we even have the gall to step foot on this continent—to return to this continent?

The students have drifted off to bed—Round River has built a stucco two-story bunkhouse, with a thatched roof—and Mike fills us in a bit on his life and his work. He and Dennis had spent a little time together before during one of Dennis's earlier visits, but most of their contact has been by phone or e-mail, and Mike talks slowly, almost carefully—not guarded, in the least, but just carefully, almost as if from some internal governance, it seems, some code of manners that is respectful and even cautious of the distance and the difference between e-mail or telephone and the real-life application of sitting by a fire under the stars, sipping tea, in Africa. As if time moves differently—slower, of course—and that because of this, the old rush of sentences might be ineffective here.

Maybe I am making too much of it. Maybe Mike's manner and bearing—this sense of quiet waiting that I perceive, as if he is balanced on something—is nothing more than the rarity of good manners, an extremely kind and generous heart, and an internal as well as external elegance that not even his scruffy bearing and attire can mask.

But I think it is both of these things—the elegance and respectful bearing like a mask or cover over that deeper core. And as such, he's far more of a listener than a talker, so that his story (as well as that of the rhinos) takes longer to unfold, or to reassemble. Or at least it seems that way to me, still psychically mired as I am up to my ankles in e-mail land, with the residue of it perhaps still visible on both Dennis and me like the radioactive aura of cobalt, or a corona of phosphorescent blue.

In visiting, we learn in time that Mike makes it back to England about once a year, for Christmas, to check in with family and old friends, and stays for a few weeks, then comes back to Africa.

"So you almost never get back?" Dennis says, and Mike looks surprised and says, "Once a year, almost all the time."

This year, he says, he won't be going back—the first year he's missed in quite a while. Maybe later in the spring, he says, though maybe not; the spring's looking pretty busy.

He still has a perfect British accent, but he's not from there anymore.

He and Dennis visit for a while about Round River's work

with SRT—the students, under Mike's tutelage, are measuring displacement behavior and approachable distances to rhinos —how far the rhinos run when spooked by humans, and how close humans can get to the rhinos under a variety of circumstances without bothering the rhino—male or female rhino, female with young, upwind, downwind, crosswind, hiding behind a tree or out in the open, on foot or in a vehicle, and so on. Such information—such protocol—will be foundational in establishing any sort of ethical and sustainable tourism program, wherein area villagers take a small group of clients into the desert, not to kill or horn-whack the rhino, but simply to look, and to then withdraw, unobserved and, hopefully, unfelt and never known, or little known.

Dennis and Mike talk with a curious mix of resignation, ambivalence, and hope about the pros and cons of Mike's continuing to pursue a doctorate degree, utilizing his ongoing research at Palmwag. The research is going to get done anyway—there's no stopping Mike on that. He'll continue to scrap and claw, fundraising little nibs here and there, to complete it—it is vital—but they both agree that the formal attachment of a PhD to his name will help the authority of his findings. And that, in turn, will help the rhinos, which will in turn help the rural villages on the outlying perimeters of the rhinos' home ranges.

The doctorate program, even if done largely by mail, will take extra time that Mike doesn't have; the days are too full, the responsibilities and tasks demanded of him are already too many —but he guesses he can get up and work on the doctorate in the middle of the night, and that anyway, the discomfort will last only for another year or two. As he once must have told himself a dozen years ago now upon first landing in the big city of Windhoek and being handed a year's worth of correspondence and clippings and told to begin filing.

There was a way out of that—straight on and dead ahead—and there will be a way out of this; it will just take another dose of extra work. One more round of extra work.

We finish our tea—it's late, suddenly—and Mike leads us to a little open-air stucco and thatched-roof pavilion with cots and mosquito netting, where we throw our gear—"kit," he calls it —and collapse into the darkness, sleeping so hard that neither of us hears the lion coughing and roaring nearby that night, which Mike, upstairs in his office, working by the stored-up light of so-lar panels, says he heard on and off shortly before morning.

Daylight at Palmwag is surely like a morning on Mars, though with life. A dry creek runs through the red-rock landscape, and from that arroyo birds are calling, busy in the little window of time before the crushing heat of the day arrives.

Farther downstream, the students have told us, only a couple of hours' walk distant, there is a sudden deep pool, a small water-fall where the water, traveling just beneath the stony red earth, spills out and fills a basin below, ideal for swimming. That's the place for bird watching, they tell us. There are turtles in the pool —not tortoises, but web-footed, aquatic desert turtles. How has such a creature been able to survive and develop in such a desert landscape, in what is clearly such a tiny universe—this pond little changed for the last seventy million years? These turtles, like the plated rhinos themselves, are not refugees or seed drift but natives too, born from these rocks, and born also, perhaps, from the one and only one wellspring of only that pool, amid the all-else heat of the brick-oven waterless landscape.

It's as if the seed for such an idea—the idea of a turtle—has resided almost always within the center of the earth, in the fire of magma and molten earth-guts, so that once that fire found or forced its way up some smokestack vertical rift or fissure (the chimney-clefts opened by the grudging drift and buckle of the continents themselves, or by the one great slab of Pangaea bending and arching before fracturing and then being carried slowly apart, slowly away), and once that outpouring of the earth's belly-fire hardened beneath the azure sky—an ancient and seemingly endless gestation—only then, undeterred, did the seed of the turtle, the *idea* of the turtle, dreamed perhaps, long ago, clamber up out of that basalt, as if all that time it had been waiting only for the fire to cool.

As if the idea of the turtle, and the idea of rhino, giraffe, wart hog—as if the seeds for all of it—had existed always down in that maelstrom, before being released finally to up those Pangaea cracks and fissures—the great hands of *something* flexing and fracturing the continent, reshaping the puzzle pieces, and then breathing that insistent breath on them, setting them in motion.

In Namibia, it all feels like Day One, and yet it—life—all feels like it has already happened a hundred times over, or a hundred hundred.

Mike works in what is surely the greatest office in the world, more of a Tarzan kind of tree house, or a Peter Pan fortress: an open-air stucco building with doorways through which gentle breezes are nurtured, and deep shade in a land of little other shade. Solar panels power the lights and computer, and there's a satellite phone on the wall. Snail mail can still take a month each way. It's more like a pueblo than anything—again, an open-air pavilion, with benches and work tables—and the bookshelves are filled neatly with fat notebooks overflowing with SRT newspaper clippings, scientific studies, research papers, and correspondence, each notebook labeled neatly: twenty years of activism in each office, like the growth rings of a tree, or like a thin cross section of geological datum.

The outside porch—again, resplendent shade amid the rising, dazzling heat—seems more like a vacation veranda than a work station, and the table, with its worn benches, more like a picnic spot than a scientific or activist battleground.

The perimeter walls of the pueblo-veranda are decorated with feathers and odd twists of desert-varnished wood, nuggets of turquoise, and glittering crystals and minerals of gemstones such as one might find on the windowsills of a boy's clubhouse. Everywhere, it seems, there is an impromptu gathering of the world's elemental things, a tiny cairn or pile, celebrating the things of the made and visible world as if such prizes, such treasures, might safely surround and protect the scientists within while they labor and muse and dream within the abstract and invisible life of the mind. As if such tiny cairns might protect them too in the life of the unknowable future. I've been in the offices of a lot of grassroots environmental groups, but this one makes me laugh out loud without even knowing why. You could get a lot of work done in a place like this—several hundred miles from any city— and surrounded by the things you love most.

On the back side of the office, facing the southern sun, again seeming in an arrangement of ceremonial repose, lies the darker fruit or residue of this beautiful land: the enormous sun-bleached skulls of nearly two dozen rhinos, lined up one next to the other next to the other, all looking south with their eyeless sockets and their truncated horns, all with bullet holes riddling the massive brainpans as neatly as if drilled there: a count, a marker, of loss, if not yet failure, skulls that the SRT patrols and anti-poaching units have discovered over the years on their expeditions into the desert, horrific reminders of the forces seeking to undo and erase the very things they seek to save.

Seeing the giant skulls stripped of their flesh and lined beneath the sun in such fashion, in such a strange and unfamiliar landscape, makes it easy to believe that such poachings, such killings, were performed by an alien species—by a creature far more savage and heartless than our own.

Maybe there was no choice, I think, forcing myself to stare at the collection of ghoulish skulls. Maybe every one of those last rhinos fell so that one more human could survive one more week. But I do not believe that. They were killed for greed and for lust, and they were killed to support one war or another. If anything, they were killed so that men might die, not live.

There are those—many—who believe that all wars, whether among man or animal, or even plant, arise from a single territorial imperative: the never-ceasing defense or acquisition of resources.

We want to believe we are more than that, and the very fact that we want to believe this, and possess this yearning for the unseen things—the secret chemistries of love, for instance—must surely disprove the one-story, the war-story, of our blind and eternal pursuit of any and every resource.

Would not such an unrelenting hunger be nothing but a hell? To emplace within us the hunger for invisible things makes no sense. Such unseen things must exist, or they would not be within us. As the land and the world dreamed a rhino, and dreamed and made mankind from out of that basalt, and out of that clay, so too must the emotions of love and hatred have existed within that stone; and thus, it would seem, within the fire that made the stone.

Those are the old rhinos, however; the ghosts of ten and twenty years past. The thinnest of the bottleneck is surely behind them, and behind us; the ninety-two guards saved the ninety-two rhinos, the dehorning appears to have at least not harmed the rhinos. The community prestige of being a rhino tracker has perhaps helped save them, by elevating the rhinos' immediate worth to an adjacent community beyond the aesthetic: and the government of Namibia feels now that they have sufficiently recovered the population to the point at which they can now relocate some rhinos into areas of their former range: a recolonization, a repatriation.

This is to be one of Mike's projects, and a rite of passage it will be, for him: shifting a large segment of his work over from the desperate conservation of rhinos to the expansion and dispersal of those same rhinos. Is he then like a little god, or a godfather, or merely an acolyte, a disciple, faithfully observing and transcribing the movements and needs of his beloved, attentive to every detail? The diet, the browsing habits, the preferred vegetation, the napping places, the times of day in which the animal moves, the estimated weight and general physical condition of each animal, and the tracks of each individual. The horns, the curious feet, and the silhouettes of the barren mountains behind the rhinos, with their names uttered by Mike like a catechism. He says their name quietly, almost more to himself than to anyone else, almost every time he looks up and sees them. Is all this work merely celebration and worship, or is it something even stronger and more muscular, like a minor partnership in if not the original creation then the ongoing one, remade each day, each year, as the earth drifts and then pulls back together, with even the continents themselves and the oceans' push and pull—much less our own lives—stretching and contracting like yarn being woven by a knitter's needles?

Dreamer and scientist, politician and acolyte, Mike has a place already picked out for the first rhino, some garden planned and prepared and even planted but not yet arisen. It's far enough away from any villages for the rhino to be comfortable, and yet there's one lonely road passing through, from the high points of which a traveler might be able to look out at the horizon with a pair of binoculars and see the faraway shining object drifting slowly across that reddened landscape. The creature so far away that the red has faded to brown and then to silver, washing out in so much light until even the earth is almost the color of the sky, and the slabs of the pale rhino's sides are catching and reflecting the sun almost like mirrors, and the rhino putting the viewer in mind of the sight of a whale moving slowly across the ocean, far out at sea, never submerging, just puttering, or of some other fanciful notion. And yet it is not fanciful; it is the real world, as made and physical as a chunk of iron, and is simply one to which you, despite your advanced age on the earth, have not been exposed, have never witnessed or even imagined. A door through which you have never, or not yet, passed.

And it's a strange idea, this notion of rounding up rhinos, placing them in the back of a truck, and ferrying them across the desert a few hundred miles, to reinsert—like some kind of gigantic and crude gene-splicing—into what was once their native habitat. It would seem that a three-thousand-pound animal capable of galloping at speeds of up to forty-five miles an hour and traveling up to three days through hell without so much as a sip of water would be supremely positioned for expanding its range, slipping powerfully through the net of whatever thin obstacles it might encounter, out in this abandoned old war-torn, windswept corner of the world, but such has not been the case.

Instead—based, at least, on the limited research to date—rhinos seem to be very poor or at least cautious dispersers, essentially as rooted and faithful to one place as might be that same chunk of iron to a magnet, which raises an interesting question: Did they arise in one place and expand inch by inch across the millennia and then the eons, creeping and spreading like the continents themselves—or did they arise as individuals, from numerous vent shafts, from numerous and similar soil chemistries and breath-whisperings, breath-summonings, around the globe? There are Java rhinos, Sumatran rhinos, white rhinos, black rhinos, pygmy rhinos . . . Were they all once colonists on the same piece of land, or in the same garden, which then broke apart and drifted, like cakes of ice, or, again, an ark, bearing on each drifting island at least a pair of rhinos, like grace-fallen castaways, to new locations, where the force of the world continued to shape and slightly modify them—or did they long ago begin to sprout up here and there around the world in numerous different generative sites, as if they might once have been as common a response to landscape and environment as, say, mushrooms after a rain?

Intelligent design is a phrase very much in the news these days, a Trojan horse for the fundamentalist and neoconservative argument for God—arguments that are becoming increasingly belligerent. Why? What God needs an argument before such a creation as the blue-green earth? Either you are dazzled and believe, or are dazzled and don't believe, or are dazzled and can't say for sure one way or the other: which, as I understand it, is of course precisely the way a God or Designer would want it, in that such an invisibility would necessitate the very faith that empowers the relationship.

Celebrating the great works of the world is easy; having faith in them and believing certain unseen premises is the stuff of religion, as well as, perhaps to a lesser degree, art. Of nearly everything, perhaps.

Out here, so much can be seen, for so far and in all directions, that it feels sometimes that the unseen things are almost visible: as if the viewer or traveler is looking out a transparent pane of glass positioned about midway to the horizon.

In the Namibian desert, then, it raises the question: If the traveler senses that such a clear-pane wall exists, perfectly spotless, with no details seeming different on one side or the other, and with the sense of that wall right there and so very near when all one's life before a traveler had never before encountered such a sense, what, then, might it feel like for the traveler to walk on ahead and pass through that clear fixture, halfway on up ahead— and what might be the consequences of such a passage?

So much remains unknown about the rhinos—as if, despite being invisible, they too are composed somehow of at least some of the elements of that same unknowingness—and so much is unknown too about the desert in which they originated, and which is now all that is left to them. The surface of the country and the thick overlying skin have been counted, cataloged, measured, and inventoried attentively by scores of geologists, enthralled by the possibility of wealth of the sort that is spewed all over so much of the rest of southern Africa—the spectacular diamonds, buried deep in glittering dikes and towers, of course, but rumored also to sometimes be lying scattered and cast about at the surface, as if having worked their way through a hole in someone's pocket or having fallen from a purse: diamonds that are believed to have been ejected in a cascading, glittering explosion, up one of those same vent shafts or fissures made by the cracking, drifting-apart earth, a fountain of diamonds being hurled far into the sky like rockets before raining back down to earth in a clatter, the cast of diamonds arrayed like an irregular net, many miles in diameter, hurled randomly across the desert.

And not just diamonds, but other gemstones—amethyst, aquamarine, garnet, topaz, tourmaline. The geologists scurried across the landscape, marking the grids on their maps—the geologists' desire itself like an imaginary net cast over the landscape—hunting. Other times they wandered in looping, circular fashion, as if lost: noses to the ground, one step at a time, searching for the invisible thing amid so much other made and physical startling splendor. Searching for wealth amid paucity.

Malachite, shattuckite, fluorite, heliodor, blue chalcedony: they existed, and still exist, in great concentrations and abundance, throughout Namibia and South Africa, underground in deep and secret vaults and caverns, and on the surface, too—gems and minerals as bright as the birds above, the crimsons and ceruleans and startling sapphires that flit through the fronds of the shade trees and thickets—weaver birds more lemon bright than any gemstone, so bright as to seem to sear the retina.

Blink, however, and the birds are gone, having eluded us. Is this, then, the allure of gemstones—that they are more graspable? Throughout this part of southern Africa, the geologists found only what they found, which was nothing of worth: old basalt, and more basalt. The cooled fires, the cold fire.

The geologists hurried on, lured by the shouts of other, greater wealth elsewhere in Africa.

One can still almost feel the echo of their speed, the alacrity of their passage, in the landscape's ringing emptiness; one can almost feel both their disappointment and desire, amidst this red nothingness.

And yet, ironically, the geologists did leave great storehouses of beauty behind, and untouched. The surface of the desert is scattered with tiny spherical specimens of varnished turquoise, blue as a bird's egg, like the tiniest of Easter prizes amid all the larger red cobblestones: turquoise balls no larger than a marble, amid the softball- and bowling-ball-size red cobbles of basalt; and here and there too are magnificent specimens of quartz crystals, spires and spindles of translucent or ivory crystal sprouting in dagger-toothed clusters, evidence of old fault lines against which the basalt had cooled, with subsurface waters carrying in steaming solution the siliceous makings of those crystals-to-come, where, cooling ever so slowly within and beneath the almost timeless overburden of so·much basalt, the quartz blossomed, accruing more and more atoms, which arranged themselves in mindless yet ordered algorithmic design . . .

The geologists had no interest in quartz, or in turquoise. They hurried on, as if blasted from a cannon. I have little doubt they will be back—they always come back—but for now there is only ringing silence.

There's always something we want, even when there is nothing. Even when there is only sand, we will want the sand, and even where there is only the wind, we will want the wind. When there is only sun, we will want the sun, and when there is only the hardened shape of the earth, we will want to carve it and reshape it into something else.

This is the plan, in fact, along the border of Namibia and Angola: to build a dam in a canyon and to convert desert to lake, and, like a miracle, to extract great power from the slack-water depths; to convert the life-giving artery of the Kunene River into yet another hydroelectricity project with the relatively short life of the dam—fifty years? The span of less than a human life will be exchanged for the ecocide, the reversal and upheaval and destruction of yet another unique ecosystem, and exchanged also for the likely destruction of yet one more indigenous people, the Himba, with their bright red-clay paint that blocks their skin from the desert sun, and their ancient culture, ten or twenty thousand years old in the world, or older.

Like all indigenous tribes, the Himba, with their painted bodies, can be said to be unique, and yet their destiny appears anything but. It is discouragingly similar to that of all native peoples who stand in the way of a dam—the dams will come, have always come—and if the world has not protected indigenous cultures from dams in the relatively mellow past, how can the Himba be expected to win—which is to say, to survive—before the all-chewing, all-gnawing beast of this century?

There can be no end to our hunger, our need, our great loneliness, and it occurs to me that before it is all over, there will be those who will circle back to this corner of the world, one of the rare places in the world where there seems to be nothing of worth to us—no oil, no coal, no diamonds, no trees, no water—nothing but basalt. No one, not even the Americans, have figured out how to chew up basalt and expel it into something "useful."

But then I remember. They—we, the world—have already been through here, have already gnawed and taken the only thing there was to take: the dried, dense hairlike sheath, the fingernail keratin, of each and every rhino's horn. It was all there was of worth, and we found it.

Contrary to some beliefs, rhino horn is not valued in the Far East as an aphrodisiac, or as a potency aid—seahorses, deer antlers, and who knows what else are used for that. (An estimated twenty million seahorses per year are dried, flash-frozen, and sold in prepackaged little envelopes around the world. As a result, in the last decade, Southeast Asia's seahorse populations have plummeted by 70 percent.) It seems there is nothing we won't consume, and with our own numbers and appetites growing, I have to wonder sometimes what will be left, not at the endgame of our species but even a more modest thirty, forty, fifty years out.

This is not to say that traditional cultural medicine, or TCM, doesn't work. Rhinoceros horn has been prescribed for life-threatening fevers and convulsions, and, according to the journal *New Science,* has been clinically shown to possess fever-reducing capabilities.

TCM is still stigmatized in the Western world, even as it has often provided the base for numerous successful "Western" medicines. "Ephedrine," reports *New Science,* prescribed in the West for asthma, "comes from a plant that has been used in Chinese medicine for millennia. Artemisinin—extracted from daisies and used by TCM practitioners for 1,500 years, is now a very promising anti-malarial drug in the West. At least a quarter of the world's population, including Chinese on all continents, Koreans and Japanese, use medical practices based on TCM. Trade in Chinese medicines was worth around $2 billion in 1994, and is growing rapidly."

The list of plants, animals, and minerals utilized in TCM now lists 11,599 different entries. Add to this an exploding demand for TCM, in part due to increased prosperity in Southeast Asia, and the effects on wildlife can be devastating.

It raises the old question of farming and captive breeding programs, as well as the legalization of once illicit substances, in order to control and regulate the trade of these things, thereby, in theory, diminishing the devastating black market trade. The British tried to legalize opium in the 1800s, going to war over China's refusal to do so.

In this view, there are some landowners in South Africa who have received permits to raise captive rhinos, or domestic rhinos, for lack of a better term — trafficking in the legalized zoo industry — and because of the slow but steady recovery, the fledgling recovery of the Namibian rhinos, there is a second plan developing, concurrent with Mike's and SRT's plans, to help recolonize their old territories.

In an era of diminishing, even disappearing budgets for the protection of endangered species and their habitats, or indeed, it seems in the United States, for almost any conservation-minded activities whatsoever, activists and advocates are called on increasingly to craft creative nontraditional partnerships that will enable them to continue to pursue and reach their goals. And when Mike tells me that another of the conservation community's ideas to raise money involves the killing of rhinos — *excess rhinos*, as they are considered, though I do not think this is Mike's assessment, and I know that it is not that of his once-upon-a-time boss, the legendary Blythe Loutit, about whom entire books could be written — I have a bleeding-heart liberal reaction, which is to exclaim, "What bullshit," even as I understand too clearly the reasons and excuses.

In a moment, a little bit — a very little bit — about the mercurial and visionary and enduring Blythe, who, having dedicated her life to helping scratch and claw out a slightly increased life expectancy for each and every remaining rhino — dreaming of their survival, and then enabling that survival, even in a time when all must have seemed hopeless, with poachers and military hooligans running roughshod across the country—is understandably loath to release, even for cash, that to which she has dedicated so much, to what she must surely perceive as the vapid disengaged perverse desire of a rich white (or any other color) hunter to shoot a bazooka at such a marvelous and complex and crafted creature as a rhino for the sole purpose of watching a powerful three-thousand-pound animal fall to the ground with a great *thump*, powerful no more.

To consider approaching an almost blind animal to within any distance whatsoever—even point-blank—only to then drop it with some great cannon of a weapon, repulses me, and my guess is that the pleasure, if that is the right word, for the so-called hunter might reside as much in the opportunity to find usage for such a howitzer as in any real relationship or engagement with the species that is at the receiving end of that howitzer.

Wouldn't you think there would be two or three folks out there in the huge sea of wealth, and in all the wide world—just two or three—who would pay just as much to see a rhino not killed?

Blythe Loutit feels the same way, arguing that the aging "excess" rhinos that would be targeted for the fundraising hunts would better serve the needs of conservationists and the protection of the species by being placed with zoos, or on private ranches engaged in education and outreach programs: captivity before death.

Working in conjunction in Namibia with the Ministry of Environment and Tourism (MET) and the Community Game Guard Program, SRT initially employed convicted poachers to protect the rhinos because of their invaluable and extensive knowledge of the landscape and of the rhinos themselves. In their book, *Heat, Dust, and Dreams,* Mary Rice and Craig Gibson re-

port that "the black rhino population in the area has all but doubled since the formation of the Trust" and that "poaching has de-
clined exponentially."

It's daunting to me to be reporting on rhinos in Namibia
without Blythe being present. The weakness of too many fledgling grassroots groups lies so often in what is also the germinating
strength and seed of their creation: the "formidable and determined" activist, the "spearhead," the "heart and soul" or "back-
bone" of the organization. An imbalance is created in the necessary and initial surge that is required to initiate change.

The founding director's force of will alone, it seems, carries
the organization for a great distance, accomplishing great good,
with the organization moving forward forcefully, like a brightly
painted sailboat catching so much available wind: *knifing* forward
across what might previously have been for so long nothing but
slack waters. As if that one person, or that person's will—desire,
or yearning, or outrage, or fear, or love, or a potent and unmappa-
ble combination of all of those emotions, and more—had been
the breath, the wind arising, like something summoned.

And sometimes—rarely, but sometimes—the resource
(rhino, or wolf, or grizzly, or old forest grove) is protected, or,
more commonly, its imminent destruction is delayed. But in the
meantime, twenty or thirty years have shot by. Or perhaps they
have crept by, in the ground-eating manner of a tank, or with the
proverbial glacier's creep, and one day, as is the nature of all things,
it is time for another to step up and step in and fill that space, or
to help fill that space.

If it sounds as if I am saying it is that time for Blythe and
her husband, Rudy, at SRT, I need to be clear: I am not. They
have never been more effective or engaged; indeed, witnessing
the success of their undaunted efforts is a fruit few activists ever

get to taste, and that success is empowering them to do further good.

All I am trying to suggest is that in my own mind, with Blythe currently recovering from a lengthy and severe illness, it seems to me that she and SRT are extremely fortunate to have found Mike, and to have spent these eleven years grooming him.

I am not trying to justify Blythe's absence from this sojourn, but rather, to excuse my absence from her, and to be looking so intently at what I am assuming is one of the bright and younger lights of SRT, particularly with regard to building lasting relationships in the communities adjacent to the rhinos' refuge, *Mike.*

The tiniest bit about Blythe, then: the iceberg beneath the tip of SRT. She came to this landscape as a painter and first fell in love with it as an artist before being summoned to conservation. That was a long time ago, and in the intervening years, and then decades, there has surely not been as much time for her paintings as she would have liked; and yet again, *look,* here are the rhinos, surely more fantastic than any representation of pigment on canvas. She married Rudy, and they set out with a handful of government officials to make a stand, laboring to convince Namibia, and the world, and—first and foremost—the native communities, that the rhinos were an asset, not a liability. That the relationship was complex and therefore powerful.

Among the other old graybeards, elders still engaged in the issue, is the highly respected Garth Owen-Smith, who served as an agricultural superintendent, as well as a conservationist for the Namibia Wildlife Trust, where, writes Mary Rice, "he was instrumental in developing grassroots community involvement in an attempt to halt the plunder of the areas' wildlife."

• • •

I ask Mike what his thoughts are about this proposed rhino shooting. He's as tactful and diplomatic as are some activists fiery and frothy, acknowledging that while it's understandably troubling to some, the rhino conservation community absolutely has to raise more money somehow, or it won't be able to expand home ranges, which stalls the recovery and places the surviving herds at greater risk in their long-term isolation. Mike shrugs a scientist's shrug, and it surprises me: I'd forgotten that. He *is* a scientist, and I admire that he can remain one, even while being an advocate as well. And I know also, from what Dennis has told me, that Mike himself has made many long journeys around the world, visiting funders, trying to scrape up enough funds to continue his research. There are triple responsibilities involved, in these days of vanishing research budgets: doing the science, doing the fundraising, and sometimes working like hell as an activist to try to help keep the very subject of one's study from completely disappearing.

It must seem a little like fighting three civil wars at once, I think, and yet Mike does not seem overwhelmed, does not seem cinder-burned, seems nowhere near the point past fatigue. He still seems bright and capable, and with his eye on the longer-term goals. I get the feeling that he too would prefer the aging rhinos not be shot like nothing more than old slaughterhouse bulls; but with his own energies and the resources of his organization stretched thin already, he possesses a veteran activist's understanding of what to get his knickers in a twist about, and what not.

I'm all for expanding partnerships between hunters and environmentalists—for decades, it has been a severely underutilized, often almost nonexistent alliance, with the environmentalists, in my view, sometimes guilty of a lack of outreach; but the hunters too have an ethical responsibility in this newly developing relationship, and just because they can now kill something they previously weren't able to doesn't mean they *have* to.

Hunting is, or should be, about choice and selectivity, and I continue to wonder if there's not a more creative and productive way to accomplish the desired end results for all parties involved. How important to the hunter, really, is the blood and the death? In the case of edible quarry, it is critical; but in the case of inedible quarry, is it not obscene?

Conservation organizations are making creative efforts to address this issue. In some African countries, they've taken to issuing darting permits, where the rich folk pay for the privilege—I can think of no other word, even as I am aware that that is not the correct one—of stalking and then shooting rhinos with tranquilizers on the occasions when management objectives call for that course of action. If such things were really important to the sport, I suppose that a photo could be made of the shooter, the darter, sitting aflank or astride his or her trophy, and that a synthetic horn could even be manufactured from a cast made of the original, though I would not hope or expect that such accommodations would soon become part of a growth industry.

But who among us has the authority to police the morals of others; and who among us, amid the sinking wreckage of a once wild, once supple and vibrant planet, has the time, particularly with regard to strategic effectiveness, to care for, much less act on behalf of, the individual anymore? The aging bull rhino, once priceless, though now, a few short years later, superfluous; the giant larch tree, or even a grove of such trees, scheduled to be bulldozed, against the greater fabric of a million acres or more that needs defending.

There are only so many fights in which you can engage, and each fight is its own sort of loss—the destruction of your own goals, ever receding, it seems, from any peace within. An activist, a laborer on behalf of rhinos, might choose, foolishly it seems, to seek to control the actions of an abstract individual in some distant country—a man or woman who desires to draw a permit to legally shoot a rhino—or to expend the same amount of energy working to protect a hundred rhinos, or five hundred, and a million acres of Damaraland, while strengthening also the human communities at the perimeters of the rhinos' range.

There is only so much time. You have to be ceaseless and enduring, but you must also get in, do the job, and get out. The trick is to help protect the rhino, or the old forest grove, or what have you, without destroying yourself.

Part II

WILD

WE HEAD OUT INTO THE FIELD — IF SUCH A DISTINCTION can be made between where we are encamped desert-midst and any farther points on the horizon, equally remote, I cannot discern it. A couple of trackers have joined us, Joseph and Leslye. The heat of the day has passed; the rhino trackers avoid the middle of the day to keep from potentially stressing the animals overmuch, and instead track only in the morning hours, and again late in the day.

During the morning sessions, they track until about noon, after first allowing the rhinos an hour or two at daybreak to get up and move, laying down their spoor, before the trackers head out farther into the desert, searching for the fresh telltale prints crossing the sandy wash of a dry riverbed, at which point the trackers leap out of the truck and follow the tracks, often across bare stone, carrying neither water nor lunch nor compass, traveling until they catch up with the rhino, at which point they mark in their journals its identifying characteristics: the horns, the ears, and the general condition of health, as per Mike's system of ranking. The trackers possess no map but carry GPS units, into which they key the coordinates of each rhino's location.

Mike and Dennis and I ride in the back of a truck, following an improbable dirt and stone road that leads surely to nowhere, the futile legacy of either exploration geologists, or a military trying to lay claim to that which no one else wanted anyway. The barren land unscrolls before us as if being created by the very act of our seeing: as if, everywhere we turn our eyes, new landscape springs into being, unfolding as if from some secret place within us. As if we have been blessed with powers of creation—was it once this way for a Creator?—and the effect is humbling and invigorating. The closest experience I have to it is only that of the imagination: it feels like what, as a child, I used to think it would be like to land on another planet—Mars comes easily to mind—and to wander around lonely but invigorated, first-seeing.

It is an entirely different quality of loneliness—crystalline, brittle, and haunting, but not painful, possessing instead great wonder and beauty. And it makes sense, I think, in the presence of such stellar, lucid loneliness, for either a Creator or the land itself to labor with extra care to fashion curious and unique responses to that loneliness: if not to fill the landscape with an overabundance of creeping, crawling, leaping life, a shoulder-to-shoulder abundance of it, then to at least work to make it, that life, extra interesting, and to fit it with extra precision into the challenging slots and crevices and opportunities—perhaps fleeting, or perhaps enduring—on that landscape. Or so it seems to this first-time traveler.

As we bounce along in the back of the truck, Mike watches us watch the landscape. He keeps an eye out for rhino sign, but he watches us too. "It's always great to see someone new come to Africa," he says. And it seems to me that in his watching, he is looking for something, a thing not yet lost to him but just now beginning to recede: the memory of how it first was for him, so long ago. The awe and freshness and delicious zest and worshipful wonder, in the long days before he became fluent in the rhythms of time and timelessness.

As if, where ten years ago might have seemed like only ten years, perhaps eleven is starting to feel like twenty or thirty, and where twenty might feel like a hundred.

As if he too is still entering another landscape unknown to him, the unknown landscape of time. Not with regrets—with anything *but* regrets—but with curiosity and wonder; and it seems to me, again, at least a little bit, that in looking back at us, and at our innocence, he is not just observing but is maybe—just a little bit—trying to remember.

We drive, and we see nothing, beautiful nothing, for a couple of hours, until finally, dark in the distance, there is a small band of oryx. So absent has the landscape been of life that at first it seems an anomaly—as if the horselike creatures, with their long spiraling horns, have escaped from somewhere else (again, the notion of another planet comes to mind)—and it is not until the oryx, also called gemsbok, break into a gallop at the sight of us, with their long tails flowing and their gait smooth and muscular, despite the cobbles through which they are galloping, that the notion of *alien* leaves our mind, disappearing as quickly as it first arose. They are instead as natural and of a place as anything ever was.

Surely there can be no other worlds like this one, and nothing else, anywhere, even remotely like an oryx, in all of the universe; and it makes the loneliness of the landscape even more wonderful, to see them running across it.

Watching them, it seems as if the loneliness is an inch-thin sheet of ice or glass, and that with each gallop they are crashing through it, breaking that loneliness into shards, and that their hoofs are flashing in the new-cut glass or ice, as are their horns;

and their eyes gleam bright in the invigorating red of the setting sun.

There can be another kind of hunting besides rhino hunting; the pursuit of Africa's myriad species of ungulates, in their incredible array of sizes, and each with a slightly different twist or curve of horn—fantastic creatures with names ending in -*bok* and -*beeste*—hartebeest, springbok, wildebeest, gemsbok—is a multibillion-dollar industry. But so too is the nonlethal pursuit of Africa's wildlife an established multibillion-dollar industry: half of the world's wildlife tourism takes place in Africa, so that the looking-at-things dwarfs the killing-of-things—and something that *does* bother Mike is the recent news that the Namibian government has issued a hunting concession to an area of land currently also under concession to one of the rhino-watching-tour-based community collaborations of the very sort that SRT has been trying to nurture.

"It's just not a good *mix*," Mike says, with his rich British talent for understatement, and I see what he means. It's not a pleasant picture to imagine the hunters driving past the wildlife-watching safari with a giant dead oryx fastened to the bonnet of the Rover, tongue hanging out, eyes crossed, flies buzzing—the very same oryx, perhaps, that the tourists might have been watching earlier that day.

The potential for unpleasantness on both sides of the fence seems infinite: the approaching jeep-load of tourists disrupting the hunter's stalk; the tourists happening upon a newly killed animal, midevisceration.

"There was no consultation, no planning," Mike says. He shrugs. "It can still be fixed," he says—there's plenty of country, all the places where the rhinos go aren't even the best wildlife habitat—but it'll take more time and energy to make the calls, and to lobby the various officials, working to undo that which was done.

In this manner, a dozen times daily, do the activist's needs and demands intrude on those of the scientists, such that the previously established control points are now vulnerable to being skewed, with the result that all subsequent statistics might be rendered meaningless. Yes, the subject rhino fled four miles instead of the predicted two, in this instance, but perhaps that was because it had heard shooting that morning, had in fact been standing next to the herd of gemsbok that were fired on. Too many variables, and the loss of a controlled or closed system.

Add to this his task of conducting his doctorate studies, and being the lead scientist as well as manager of SRT's Palmwag field station, while incorporating the Round River student volunteers, while continuing his community outreach, while rallying to travel to strategic fundraising events to explain SRT's goals, and to discuss this surreal landscape and the surreal species he follows —on top of this, Mike has now blocked out five days to ferry a journalist and an educator/fellow scientist around the desert, telling us the story all over again, from scratch, as he must have told hundreds of people, over the years. *You track rhinos? How interesting. What's it like? Tell me about it.*

What's amazing to me, riding in the back of the stiff-shocked truck with him, is how vital and enthusiastic Mike remains despite the obvious workload; how clear-eyed and patient, enduring. His energy is not the frenetic, jangled drive-on-through-to-the-end demeanor of so many activists whom one might perceive at first to be tireless (when in effect what one might be witnessing instead is a recklessness). Instead, his seems to be of a different quality.

It seems possible to me that his bonhomie comes from both

a deeper source—a happy and engaged childhood, perhaps, I think, remembering the clubhouse/tree fort feel of his office—and some incredibly wise and mature source too: as if he possesses a quality of breadth similar to the landscape itself.

I guess the closest I can come to describing it is to say that this quality of his makes me think again of how he waited for a year down in Windhoek, closer to the desert but not yet in it, filing papers. A year's worth of filing, with each sheet of paper bringing him that much closer to his ultimate rendezvous: grain by grain of time.

We have to grip the iron rails of the truck with both hands to keep from being pitched out, though Joseph and Leslye are not driving fast. Dennis and I scan the horizon with erratic attempts at focus—*groping* for the sight of a rhino is what it feels like, rather than sweeping or combing the land with our eyes—and Mike keeps on talking to us, trying to fill us in on the story of rhinos in Namibia, and on the nature of his work, and he seems to be only occasionally glancing around, doing his real work.

This is another of his endearing qualities—his ability to be fully present, whether listening or speaking, when in the presence of others, a diplomat's grace, so uncommon in the world as to seem surprising now, when encountered, and yet really, it is nothing more than good manners and attentiveness.

Despite his attention to us, however, Mike has still been looking—watching the horizon like a sailor from a crow's nest, and glancing down at the gravel lane over which we are bumping—and once we are about a half-hour out of camp, he bangs on the roof of the truck with the palm of his hand and calls out something in dialect to Joseph and Leslye, who stop the truck and then back up even as Mike is jumping out and walking off into the desert, head down, inspecting the land and stones as might the geologists who helped make this road have walked similarly, searching for clues of a less temporal nature than the shifting-sand sketchings of rhino feet. It occurs to me that the rhinos have been making these tracks in this same place with a regularity and endurance that might exceed the age of certain minerals, and certain gems. What an odd thought, that flesh and bone might in this fashion exceed and prove to be more immortal than any stone or ore.

Joseph and Leslye hop out, leaving the doors open and the engine running—it pleases me that despite having viewed countless tracks, they are still enthusiastic about the sighting of yet one more—and the three men cast back and forth across the road, circling the tracks and studying them intently, though with slight traces of disappointment on their faces, so that although I cannot yet discern the tracks, I understand that the trackers are thinking these are old tracks, maybe laid down this morning, so that the rhino or rhinos in question could be thirty miles away by this point.

The trackers follow the tracks (still indiscernible to me) in a line in one direction, then cast back, milling like hounds trying to untangle a scent: as if the rhinos themselves, upon intersecting the road, had grown nervous or confused. From time to time the trackers lift their heads and look out at the horizon.

But there are no rhinos in sight, and neither is there any place on the terrain where they might be hiding; and with only an hour of light left in the day, the trackers feel defeated by that cruel confluence of too much space and not enough time.

They circle back to the road, where Mike points out the tracks to us. He walks us a short distance out into the desert to find a good clean print in the red sand.

Looking down at it, I'm reminded of what it's like to peer

through a microscope, of the sudden disorder and rearrangement of scale, and it occurs to me, for the first time in my life, that I really don't have any idea of the size or shape of a rhino's foot.

The scuff marks in the sand are evident—it's a clear track, and I crouch and study it and try to parse it out, as Joseph and Leslye and Mike had been doing. I look to the east. "I guess it came from that direction?" I say.

Mike clears his throat and corrects me, tells me that what I am looking at is only the print of one of the creature's three toes. That actually, it appears to have come from the west.

Now the print of the other two toes leaps into focus, as does the plate-size marking of the main foot—the lone toe print I'd been examining merely the size of a man's fist—and I have to laugh at the dizziness in my mind, and the reordering of things. "Right," I say, joking, "I knew that."

As best as they can tell, the markings are those of a cow and calf. They're not sure if these are animals they've already counted or not, and again, there's a mild sense of disappointment; without visual identification, the data can't be logged in and is as unusable to the study as if the sands of the day had already blown in and across the tracks: as if the tracks, and the creatures that laid them down, had never been.

It is not that way at all—there will always be another day, and in the morning, the trackers will have another chance—but there in the red-dimming fast rush of night, it is a feeling of loneliness, if not quite loss. Again, there is the troubling sensation that time has run out.

It seems an incongruous, even ridiculous notion, here in the one place where time has barely moved at all—but it lingers nonetheless, on our drive back into the research station, and I feel lost, stranded among so much of my own unknowing, even while

86

remembering that this is one of the reasons I came here: for the privilege of reentering the world with almost complete unknowingness, and to do so, hopefully, with hunger and intensity; to enter a land of giants, which are no less gigantic or dramatic for their not being seen.

We make one last stop on the way in. How to imagine things you've never seen, in a land where you've never been? A traveler can ask of a guide all the questions he or she wants, but with no experience against which to anchor such questions, any answers may seem meaningless.

We get out and walk up to a little ridge to search one last time that day, to try to spy not just tracks but the animal itself.

The heat has left the land, so that it's cool riding in the back; already, the day's almost unbearable heat seems hard to recall. The desert appears to be swimming in a red light that is almost mercuric in its density now, as more and more violet inky-darkness is stirred in, like the palette of a painter seeking to obscure all of the day's works. Glints of gold and ocher and yellow and valiant streaks of red still shine here and there within the advancing river of the deeper redness, so that any animals we see will appear now to be wading in a river of the last red light flowing across the desert floor, with so much darkness onrushing.

There are no rhinos visible in that lake of golden-red darkness, but as we get out and walk and search and hunt, the loneliness vanishes like mist burning off in sunlight, and the imagination kindles, then flares.

As well, there are clues now that more than hint of their existence to a believer. The gnawed and frazzled branches of acrid-scented *Euphorbia* bushes: something chewed on them recently, and so toxic are the bushes that almost surely it could only have been the rhinos, having had fifty million years to adjust and accommodate themselves to the *Euphorbia* poison, and with the rhinos' great size, perhaps, somehow helping in that accommodation, that ability to withstand such toxicity. (Occasionally, kudu will nose around in the *Euphorbia* also; but it is rhinos that accomplish the mass consumption of the plant.)

There are clues—more than clues, *evidence*—in three-toed tracks and other spoor, the dung the size of footballs, and in the etchings and furrows in the sand from where the rhinos have dragged their hind feet, as if laying down a line in the sand not only to define and advertise their territory to other rhinos, but even to the believers, the hunters in search of a clue. Even the acrid scent of their *Euphorbia*-concentrated piss—an odor of almost industrial toxicity—proves to an observer that they are here, even when they are not here.

And how much of the rest of the world is this way: completely unseen, but completely present, just beyond the short lines of our sight and knowledge? A short walk on the landscape —the scuff and scrape of gravel underfoot, and the felt presence of our place between earth and sky, the senses more aware now of our own lives and bodies swimming through and in and amid that same onrushing darkness—should seem terribly lonely; and yet, paradoxically, it is not.

The landscape, felt, now labors to shift into focus. Questions come like birds in the desert alighting near a watering hole. Why does the oryx, and so many other of the desert ungulates, possess the wild charcoal stripes and smears around their face, mask-like? Perhaps it is to partially absorb and break up the desert's fierce glare so that they might be able to see farther, and more clearly, without squinting: the oryx's brand of desert-fashioned sunglasses.

Maybe every facial marking, slightly different among each species, is adjusted across time, by some painter, to most perfectly match the type and reflection of sun that each creature is most likely to experience in its habitat. Tilt the most-commonly-experienced sun from a right angle in a small animal able to withstand the midday heat to a slightly different cant for a larger, more crepuscular animal, and maybe the stripes and bands around the eyes fall and cant accordingly, like shadows.

Why is each horn shaped and ribbed and spiraled the way it is? Do the horns radiate a little extra heat, with that increased surface area? The antlers of North American ungulates (shed and then regrown annually, unlike horns, which are not shed), particularly in cold weather, are relatively slick, lacking the spirals: *why?* Even the annular growth of horns in North American species— bighorn sheep, for instance—do not possess the extra degree of spiraling seen in African horns.

Doesn't almost everything in this part of Africa have to do with the obvious dominant presence, *heat,* and, in the desert, with its twin, the near absence of water?

Or perhaps the spiraled horns, with increased surface area, do act as an extra heat dump, but perhaps they also do significant extra damage to attackers—lion, hyena, leopard, or even another of their own kind—by spiraling into the flesh, entering in the manner of an archer's arrow, designed to do the most damage possible in the one thrust.

More questions, more thinking, more walking; less loneliness. Night falls, but we are alive for one more day—it rolls in over us as we return to the truck. It seems as important as it is impossible to never run out of questions. I cannot remember being in such a state of perpetual wonder since childhood, and again, I am overwhelmed by this gift. This kind of childhood wonder has been gone for a long time.

It occurs to me to wonder how much in my own country I might or must have passed by, wondering at what point the torrent of questions slowed to a trickle and then to silence, and then to an artificial or perceived familiarity, which is almost surely not a familiarity at all—certainly not the familiarity of fifty million years, or even of forty-six years—but instead merely a comfortable constriction of boundaries: a tapering and diminishing, and the reestablishment of borders and boundaries to a more manageable, controllable, personal size, such as those contained within the individual cell of a honeycomb.

It's one of the things we do best, walling ourselves off from the often frightful outside world. Whether it is safer or more manageable within those honeycombs in the long term remains to be seen. But almost for certain, it's lonelier within those tighter borders: the walls eventually closing in so tightly, I imagine that one day there will not even be room to turn around.

The walls will be touching each shoulder, and pressing against one's back and chest. The ultimate safety will have been reached, the safety of sleep, and there will be no more questions, or even the opportunity for questions.

How fortunate we are to have Mike, for this handful of days! The next morning he loads us into the Land Cruiser and takes us back out into the desert, following one seemingly endless road. We drive for hours, and there are no other humans between us and the horizon, only red-rock and black-rock desert. Mike drives carefully and stops often: stopping at almost every new desert flower he spies, isolate or nearly isolate amid so much heat and rock and, I have to say it, *nothingness*, beautiful nothingness.

Sometimes he remains in the truck, leaning over the steering wheel and gazing at the lone flower, or the little colony of flowers, as he narrates about it, lost in his own wonder, though usually he gets out of the truck and walks out into the desert and crouches by the blossom, repeats its name for us, and implores us to smell it.

Some of the blossoms are sweet-scented, though one smells like rotten meat: a clever way, in a land of great mortality, to compete with death, attracting pollinators and spreading life. The rainy season, with its few hundredths of an inch, has come and gone, and the flowering plants have seized their opportunity, have burst forth in minute unrestrained glory, as if unaware of the immense fabric of red that sprawls, dwarfs, absorbs them.

They blaze within their own space as if this desert is imaginary: as if there is nothing else in the world now but cobalt, magenta, or periwinkle. As if they have remade the entire world rather than merely the space within the confines of their petals, and only that for another week.

Mike crouches by nearly every one, and stares at some of them for a long time. Namaqua kuni bush, short-thorn pomegranate, Welwitschia: we stop and crouch at each one, watching it quiver in the drying wind, vibrating wildly, as alive as anything on this landscape; and at each one, we remain a little longer, the three of us hunkered around it like acolytes.

I can imagine no landscape, no life, that could be further and farther from England's dewy pastoral; and when I ask Mike how he got here—whether it was the landscape or the single species, *rhino*, that drew him, he laughs.

"There was a famous zookeeper named John Aspinall," he says. John Aspinall was the most flamboyant and adventuresome zookeeper he could imagine, and an ethicist, going to great extremes to keep his various animals happy and well fed. Whereas most zoos would feed their animals any old refuse they could scrounge, in order to hold down costs — the tossed-out leavings of grocery store trash bins; browning, speckled lettuce; curled-up avocado skins; coffee grounds; onion peels — John Aspinall served his animals only Grade A-1 food, shopping for it daily and choosing only the best; and he would import each species' native foods as well, as often as he could, with the fresh crates and boxes coming in daily, as I imagine it, from the four corners of the world.

And John Aspinall's zoo, it turned out, was right in the suburban neighborhood of Mike Hearn's childhood in Kent: the Port Lympne Wild Animal Park, as it turns out, was just a few houses down, so that each night as he lay in bed, he could hear coming from just beyond the rose garden the coughs and roars of the lions, the screech and chatter and wail of howler monkeys, the cawing of parrots and toucans, the brayed trumpeting of elephants, and the subsonic grunting of the rhinos.

A different language began to fill Mike.

Mike's father knew John Aspinall from the subway rides, and visited with him at the pub occasionally as well. Mike says that early on both his parents realized that Mike needed something "a little different."

His parents could have encouraged, steered, and pressured Mike to become anything — diplomat, banker, physician, attorney. Instead, Mike's father pulled John Aspinall aside one day and asked if there might be an opportunity for a young boy to hang out at the zoo and help in any way he could. John Aspinall said yes, and Mike went straight to the rhinos.

What *is* a rhino, and just as there are different trophic levels within and upon a landscape — and just as such layers exist in our own lives as we proceed through them—are there possibly similar trophic levels of other lives, and other dreams?

After the old forest disappears, we assume that the grassland returns in its place, and that the cycle begins again. But what if there are other cycles beyond the one we perceive to be the furthest—and what if there are other layers, too, stacked above us as well as below, in which the energy levels of life are stacked so high that they reside beyond our view?

Geology, weather, hydrology—there are more braids than we can know, and somehow, beyond what anyone might ever have imagined, they have conspired to create a rhino, and not by accidental happenstance, but surely by at least some hint of planned design. I do not see how fifty million years of such over-the-top specificity can be accidental happenstance, but must instead be a marker of some kind, a trail leading either back to the world from which we came, or to the one into which we will be traveling. The rhino as gatekeeper of great beauty.

How could any child residing next to that gate—indeed, almost any adult, but certainly, almost any boy—not choose to approach that gate?

In the early days of ecology, a pyramid of life was envisioned, in which the most efficient utilizers of the sun's energy—the single-celled organisms and the cheerful photosynthesizers—formed the vast mass of the pyramid's base, with a subsequent tenfold decrease in efficiency as one moved higher up in the pyramid, converging toward the pinnacle (and beyond that, the invisibility).

The eaters of grass occupied the next level (with decreased world space available to them, due to the converging and diminishing space within the upper reaches of the pyramid), and correspondingly, the eaters of the eaters of grass had even less space—

were rare in the world, with their lives and habits being so much more expensive to maintain, from a calorie and energetic standpoint.

At the very top of the pyramid, then, were the eaters of the eaters of the eaters of grass—not just carnivores, but carnivores that would eat other carnivores—and beyond that, well, maybe the system falls apart, we're not sure.

Most problematic of all, with regard to our own relationship to the world, is the fact that we have learned to consume without eating, placing us, perhaps, beyond the pinnacle or apex of that pyramid, and into the ether, the dream, that lies beyond—and the farther into that space or ether we travel or float, the lonelier it might become.

Which is somehow, I think, part of the reason why we are so compelled to stand and stare at rhinos: recognizing at some level their own improbable habitation beyond the peak—their hugely expensive presence in the world, yet one that is somehow negotiated, crafted, blessed.

In such a model, it was assumed in those earlier days of science, the pinnacle species (despite often being fierce and seemingly dominant in their native landscape) were also extraordinarily dependent, relying upon the largesse of biomass beneath them. Take away the footings or foundation of the pyramid—the rotifers, diatoms, lichens, krill—and the structure leans, wobbles, or tilts, so that the top is no longer the top.

And though there is truth to that assessment, there were other truths that were not yet seen. Recently, scientists have been discovering that the relationships from bottom to top of the pyramid are not as linear or one-sided as initially observed: that those regal pinnacle species do not merely reside at the top of the heap like passive royalty, taxing all the peasants below with a percentage of their flesh, but that they help shape and organize and distribute and otherwise keep connected the order, spacing, and vitality of the pyramid's vast base.

Everywhere, dramatic examples are coming to light: examples that in retrospect should have been obvious to our scientists, and probably often were obvious to traditional knowledge. It's theorized now that the ancient giant cedar forests of British Columbia's coast, for instance, are in large part a function of the gardening of bears.

During the salmon runs of summer, the bears gorge on the dying salmon, which are loaded with marine nitrogen—each fish a dense torpedo of protein. Every bear consumes thousands of pounds of this incredibly rich protein and then disseminates it farther into the forest. The granite outcroppings of the coastline lack nitrogen and other critical soil-building components, so the cedars would probably not be able to survive, much less grow so large, without the grizzlies transporting the nutrients from those salmon farther into the mountains.

Nor would the little murrelets then be able to nest in those giant trees, two hundred feet above the ground, up in the clouds, nor would the secret and valuable commerce of soil proceed below, perched atop the stony skullcap of the otherwise barren granite. In this regard, the grizzly is like an active rather than passive royalty, taking life but also redistributing it, and, if not quite creating it anew, then acting certainly as the right hand of the force that has designed it, and which is creating it: the grizzly, and other apex species, seeming perhaps chosen, in that regard, and diligent and unfailing in the honored execution of that critical assignation.

But perhaps even now we don't have it quite right; perhaps even now there are still other truths, supplemental truths, unrealized or unobserved, residing just a little farther on.

What if there is a place in the model for a species that resides—almost as if floating—just beyond the known or visible town of the pyramid? Not quite the Creator or Original Designer—but closer? A species operating—surviving—on some other plane or layer of logic?

The rhino, for instance: three thousand pounds of muscle walking around on a stony, nearly barren land, one nearsighted day after another, with tens of millions of years unscrolling behind it. As if the unchanging land itself, the basalt plains, had spent twenty million years cogitating on the first rhino—had cogitated for the first rhino-less twenty million years, dreaming the perfect dream, before allowing the rhino to swell farther from the land.

Is a rhino wasteful and extravagant? Absolutely: or so it would seem, to a traditional or conventional observer of the world's trim pyramids. I think that a rhino would be extravagant anywhere it was put down or anywhere it was summoned; but all the more so for such extravagance to reside on one of the most inhospitable landscapes in this world.

Again, this disparity seems to me to be one of those rare, shouted exclamations of—what? A creator? A design? An ache, a yearning, for the miraculous?

I can hear the shout, I can see the rhino standing at the gate, but I do not know how to pass through it.

And how long did the world wait, dreaming of us, before we were summoned—and will we be a predestined, nearly eternal fixture of the fifty-million-year variety, or are we residue, experiment, like the briefly flashing ingots thrown by the steel wheels of a locomotive as it roars down the steel rails?

Hot, even incandescent, midflight, but cooling, even now, already.

Before meeting Mike, I would have asserted the opinion that landscape is a determinant so powerful as to help shape not just spirit or identity, but something even deeper.

And before I met Mike, I would have asserted the opinion that a rhino eating lettuce in a suburban zoo in Kent was not a rhino: that only a rhino, wild and free in Damaraland, was a rhino.

But even in those Kent-bound animals, there was still evidently something that drew Mike, not just from his home to the zoo a few houses away, but then, like the shadow of a bird passing in front of the sun, from Kent to Namibia. In that suburban enclosure, he had seen something, encountered something, that was clearly still present in those long-captive rhinos: something powerful enough to propel him to their native landscape, even if they themselves could no longer make the journey.

I suspect that John Aspinall might have felt slightly the same way, despite the incongruity of his position—laboring to mitigate that unfortunate estrangement, the wearing-down and wearing-away of rhino-essence.

The Port Lympne zoo was different, and even his handlers were "different." Mike says that when he was a child, "They were always the ones who were getting gored or trampled." They sought, encountered, attracted danger: "If you read about something like that in the newspaper," he says, "a mauling, a mortality" —it was always one of John Aspinall's guys. There was something different about them.

Clearly, Mike's a gentle guy—I remember the considerate silences around the campfire, in which he lets the students talk, and is content to smile and listen, no need to showboat—and already I have seen the attention and respect he gives to anyone who comes in contact with him—have seen him, already, wax rhapsodic about the sight and scent of the tiniest and most fragile of desert flowers—and so it is with a kind of amusement that I remember also some of the photographs I've seen that have been taken by Mike, photos of extremely pissed-off rhinos engaged in a point-blank staredown, or worse: the rhino with its head lowered, horns athrust, running straight toward the photographer, all four feet airborne as the rhino bends its way through the rocks and boulders that the photographer, Mike, had assumed would protect him. The rhino doing the one thing it could do best, charge: converting fifty million years of power into a single, focused moment, the intent of which was to vanquish the instigator of its alarm.

Obviously, Mike had escaped with his life each time. Less clear, however, is a point that comes out only after some more direct questioning, such as, *What were you doing so close to that one rhino?*—an especially dramatic charge photo.

Mike admits that he had been watching that rhino for quite a while that day, photographing her, and that he had a feeling she might make a charge. This is the first hint I get of a Mike beyond the long-haired peace-and-groove flower-sniffing Mike, and I press further, asking patronizing, almost parental questions such as, *Well if you knew she might be the kind likely to charge, what were you doing so close to her?*

It turns out he had an escape route planned, up in those rocks —if only he could get to them.

"So did she stop?" I ask, studying that one picture. "Was it a bluff charge?"

"No," he answers, "it wasn't a bluff charge." A little more explanation, then, seeing that I wasn't going to quit. "I had to take the pictures while I was running backwards," he said. "I was able to stay just ahead of her."

Something about his diction floors me. Perhaps it was the calculation and confidence, the familiarity, of "was able." That, coupled with the fact that he'd been snapping pictures in the midst of that life-or-death retreat, and not just rough-framed point-and-shoots, but focused compositions that captured the beauty of the animal's athleticism as well as the drama of the moment: the emotion of the moment, which in the photo appears for all intents and purposes to have been unadulterated fury.

To willingly engage such a force—confident in one's own athleticism and cunning resources—this bespeaks a person different from the one I thought I'd been seeing: or rather, another, secret facet.

"I was able to keep climbing up on rocks, and hiding behind rocks," Mike says. "After a while she gave up."

I study him—the quiet matter-of-factness with which he has related the saga of the photograph. Maybe just the slightest bit of mirth, or devilishness, just the faintest hint of acknowledgment that what he had done was beyond the norm, extraordinary. And I wonder: to whom should I direct future questions? To Mike the doctor-in-training, or to Mike the fundraising SRT ambassador, or to Mike the mountain biking surfer dude who counts coup on rhinos, or to Mike of the peaceful Buddhist monk countenance?

The latter is the one I seem to see in him most often, and the one I'm most comfortable with. But still, there's that photograph, and it's good, in a way, to know the story behind it.

It's not so complicated, then, to figure out how a life in the suburbs, when injected with just a single improbable thread of the wild, conspire to make a one-in-a-million individual like Mike.

Less comprehensible, however, are the forces that took the heat and dust and wind and aridity of the Namibian desert and swirled them together in a quick breath so many millions of years ago to weave or sculpt from that dust and wind a creature as fantastic as a rhino.

Perhaps the rhino is one of nature's all-or-nothing gambits for life to survive on a landscape. There must be efficient methods for the distribution, and redistribution, of nutrients. Small movements and gestures can accomplish this, but so too can the grand. The magnificent grizzlies redistributing the marine nitrogen of the salmon, the elephants distributing the seedpods of the Bala-nites tree—entire forest ecosystems depending on the paths and movements of elephants—and the garish parrots of the tropics, similarly ferrying the fate and future of whole forests from one location to another, as bold and bright and beautiful, as ostenta-tious, as any shouted message of man.

The forest fires that sweep across millions of acres, releasing vast new stores of energy; the churning floods that level the very forests whose resurrection the floods simultaneously summon, with their massive redistribution of fine sediments and organic matter: almost everything that is huge or powerful or dramatic has a critical purpose in nature, a function beyond flamboyance; and although rhinos may be shadows of some further and farther world of the imagination, so too might they be the single most efficient invention for the distribution of the desert's supremely limited resources: an ever-shuffling wildcard, able to trump, in three-day journeys, the great distances between scarce water, around which the smaller cycles of redistribution typically occur.

The dominant vegetation on this sun-blistered volcanic wa-terless red sheet of stone is the *Euphorbia*, too toxic to live any-where else, and too toxic to be of much use to the rest of the living world: a dome-shaped multibranched dusky blue-green bush, isolate beneath the sun; not quite a tree, but the only plant of any size that seems to grow out in Damaraland, beyond the reach of water.

A challenge and a question, then: how to break down and distribute and recycle the organic compounds gathered or summoned by the *Euphorbia*—residing midslope amid that broiling landscape like a giant fruit that nothing can eat without dying—at a pace sufficient to sustain the bright cycles of life, rather than the slower redistribution of nutrients brought about by mechanical forces, the mindless geological forces of weathering?

How to inject that deeper spark of life into the system, and what kind of maker would create a fruit that could not at least be partially utilized by *something*?

The ceaseless mowing, mulching, and munching tanklike progression of armor-clad rhinos might be just the thing called for to accomplish that transfer of nutrients across the desert, breaking down the toxins in the rhinos' magnificently broiling livers and shitting out mountains of fiber, now detoxified and accessible to the desert's "lesser" citizens, and to the thin and meeker skein of life overlying that desert.

Tens of thousands of rhinos, and then hundreds of thousands.

So they are not merely ornamentation, or spiritual representatives of a world beyond; surely they are pragmatic, too, functional and requisite in the everyday miraculous business of life.

What might be the everyday pragmatic consequences of their complete absence one day? So what, really, if all life ultimately left Damaraland? So what if the *Euphorbia* multiplied until they were a thicket overlying the red cobble landscape, and then received a spark of friction, or a ray of sunlight magnifying through a certain quartz crystal to smolder upon and then ignite a *Euphor-*

bia bush, setting the entire landscape into a horizontal river of fire and toxic fumes, one apocalyptic maelstrom—so what? Do we not still have other landscapes, other ecosystems? What would be the loss, really, of one little ecological system, or one more little universe, here on earth?

Mike and Dennis and I get out of the jeep and walk for a couple of hours. Mike points out some old tracks, some old dung, some old frazzle-chewed ends of a *Euphorbia*. The rhinos are out here, somewhere, they're just two or three days away right now—the *Euphorbia* bushes have about them the feel of a neighborhood bus stop beside which no passengers are currently standing, with perhaps not even any schedule running that day: empty now, but with the promise of some unknown return.

We're looking for tracks, scanning the ground before us and placing our steps carefully between the rounded basalt balls stretching before us to the horizon—as if someone has spilled a giant bag of similarly sized balls onto this flat surface and they have rolled all the way to that horizon, and then over the edge —though from time to time we lift our heads and look farther, searching not just for tracks, but for the rhinos themselves.

As a geologist, however, I can't help but find my attentions diverted sometimes by the sheer beauty of the rocks: the earth, the subterranean fire unleashed and then long ago cooled. Magnificent flowerings of quartz, nuggets and dagger-teethed crystals, are spread everywhere, and I cannot refrain from examining them and carrying them for a while, like a hoarder; and certain crystals I cannot resist loading into my pockets, cannot bear parting from such beauty, suddenly graspable.

Soon my pockets are filled, my shorts are sagging like a low-rider's, crystals of quartz are knocking around my knees, and I must stop and unload. It's foolish, and yet no sooner have I jettisoned the beauty than I begin again, my eyes drawn unavoidably toward the perfect crystal faces, as well as to the odd, the atypical, the uniquely irregular. How powerful this impulse is, this attraction, and how strange our definitions of beauty. Some are culturally shaped, no doubt, while others, it seems, may be more archetypal, ingrained somehow at a deeper, perhaps even cellular level, with certain assemblages and repetitions of order and form alternately soothing or stimulating us, dazzling or betrancing us, like the stars at night, or the wavering light of the campfire.

The visual allure of raw gold, for instance, or of a diamond — it would be hard not to acknowledge a certain beauty in these objects — and yet is not a rhino even rarer, and more beautiful?

The geologists have already been through here, Mike says. "De Beers sent prospectors all through this country, right after World War Two." They found no mineral wealth whatsoever — *lucky rhinos;* or lucky for a while, at least, until the next war — but even that mineral absence has not been entirely sufficient to protect the land from a different sort of mineral exploration.

The two opposing armies, the U.S.-backed South Africans and the communist-backed Angolans, seesawed back and forth in this borderland, killing rhinos to raise funds for their war efforts, smuggling rhino horn and diamonds out of the territory to fund their wars, our wars — and even now, Mike says, with the world diamond trade having established certain sanctions against the "blood diamonds" that are produced by the murderous slave labor in the mines of Angola, such diamonds are still flowing up and out of that barbarous system, and out into the larger world. They flow as if from a hellish fissure and toward our rings and necklaces, our brooches and pendants, as if drawn by some foredestined magnetism, something deeply flawed within us.

Because the trade of diamonds from Angola is prohibited, smugglers are bringing diamonds over into Namibia and casting them down into quarries, or onto the desert floor itself, and then "discovering" them and selling them in that manner, in a trade, a trafficking, as bloody as that of any drug cartel, and all for wretched beauty, curious beauty of the sort that I am spying while on our walk, idly bending over and picking up.

"The greatest prediction of rhino density," says Mike, "is the presence of rivers." And it is along these often dry river courses that seed distribution occurs, pulsing in the occasional flash floods of November—the cauterized shell of basalt-skin conspiring to channel that inch, or two inches, or even three inches, all falling within a week or ten days—into conjoining sheets of runoff that funnel down off the landscape's carapace and go rushing into the dry washes, the sand rivers along which the desert trees huddle, waiting: particularly the stunning live-oak-shaped acacia, each tree an oasis of shade and food, as well as an indicator of possible underground water.

Again, however, the magical component is the creature that is capable of transporting those riparian nutrients horizontally across the desert, like a valiant packer, rather than being limited to the dimensions of rivers—the rhinos able, as if weaving with thread, to cross from one river system to the next, and in that manner keeping the landscape and its thin supply of nutrients connected and fertile.

Water is everything. In good wet years—two, three inches of rain—everything prospers, including the rhinos. In drought years —zero, a half inch, or an inch of rain—everything suffers. How amazing is the knife-edge of existence here, the genius of creation, calibrated as it is down to that half-inch: and what buffoons we are, to think that with our massive dams and global warming and well-drillings we can patch or aid or redesign or improve such a massive and intricate system.

In such a harsh environment it seems that always there is a fork in the path, with one branch leading to success and the other to failure; and in a demanding landscape, there are more forks in the path, more choices demanded—more gambles—and hence, the ultimate survivors are often baroque: highly specialized in their success, intricately fitted, lock-and-key, to the demands of those many paths that conspired to weave one path. The harsher that path and those choices, the more ornate the solution, *survival*.

The rhino *is* a king, a victor of time, in such a landscape, as is a giraffe, and a zebra, and a lion—as is anything that has found a way to survive, much less prosper, in this land, but it seems critical for us to remember that their victory is not one of domination, but rather, accommodation.

And it feels to me that already, with our own species, so new in the world, we have chosen the wrong fork at one of the earliest and simplest of junctures—choosing to needlessly dominate and vanquish and destroy elements of the world that would better have behooved us to sustain and protect.

In a soft and forgiving world, I like to imagine that many of the choices are six of one, half a dozen of the other, that there is forgiveness for missed opportunities, or uninformed or even plain inaccurate choices, unwise experiments.

In places like Damaraland, however, it seems that the consequences are five of one and seven of another; or that they might be skewed even further, and that once made, there can be no going back, no backtracking to that earlier fork in the path.

Mike tells us that some of the species out here—birds as well as mammals, and, for all we know, reptiles too—have gambled on their ability to survive a drought by developing extra-long gestation periods, with their bodies waiting until sufficient weight has been reached to carry the embryo safely through and across those harsh times.

Others choose a different tack: waiting and watching, as if peering through a window, until there is even the tiniest pulse of an opportunity—a lush rain—at which point they rush into a full and yet abbreviated reproductive cycle, with gestation as compressed as possible, as if in a hurry to get their progeny out onto the landscape while the fruits of that bounty still linger.

The long-gestation gambler hedges her bets, waits to see if the drought is truly broken, cost-averages her resources over a longer period of time, spreading risk (though placing greater cost on her own body, across that continuum of space and time), while the flash-in-the-pan mother—if she bets right—delivers her young into a landscape in which, even if only briefly, there might still be a remnant breeze at their back.

And in such high-stakes gambles, all of a sudden the five-to-seven odds, or seven-to-five, matter immensely.

In no environment, harsh or lush, is the future guaranteed; in all of wild nature, it is the practice to conserve resources until the time best suited for their expenditure, and then to choose, in that moment of dispensation, profligacy or restraint.

Even in the productive temperate and tropical regions, the various unique markings on birds' eggs—the blue eggshells of robins, the buff and charcoal specklings of plovers and killdeer, the red speckled deciduous dapplings of wren eggs—are not applied to those eggs until the final forty-eight hours before being laid. The simple ivory-colored egg spins in its descent, still within the mother-to-be, being "painted" chemically only at the last in-

stant, with the bird mother in that manner not only able to preserve those chemicals and nutrients and resources until that very
last phase, when imminent success or failure is better able to be
gauged—when the foreknowledge of the outcome of the bet is
so close now to be almost a sure thing, one way or the other—
but able, even, to hedge the bet, fashioning an even more secure
environment for its eggs by fine-tuning the oviduct painting by
minute degrees, perhaps, to adjust for on-the-ground changing
conditions: a lush spring requiring a slight variation in egg tone
for better camouflage, for instance, compared to the bone-white
brilliance of a drought-stricken landscape.

Everywhere, and in all moments, nature is waiting and watching, and in a wild and healthy and still connected, still functioning
landscape, the traveler can feel him- or herself stepping into this
greater net or web of awareness even without necessarily understanding it. There is a certain feeling when you are in its presence,
and a certain loneliness when it is absent; and in a place like Damaraland, you can feel alone in the world of man even as you feel
assured by the surrounding world of a larger and wilder nature.

This loneliness and connectedness is an unusual combination,
and I wonder to myself what its effects have been on Mike's soul
over these last dozen years, and where, for each of us, the ideal
middle ground exists, and of the difference between our lives
now, in this regard, as compared to those of our ancestors from
even only a hundred and two hundred years ago, three hundred
years ago, when they lived in a world in which the two realms
were not cleaved apart, and in which there was less loneliness
within clan and community. A time and place, too, where those
intangible feelings of existing with a more connected web of life
also still existed.

It is almost as if these days we have to go chasing after one or the other. It's almost as if, with regard to the sustainability of the human soul, we are entering slowly into our own harsher, higher-staked environment: one of those five-to-seven situations.

It's an incredible landscape out here, addictive and wondrous and breathtaking. And yet: wouldn't a person get a little *lonely*, sometimes?

We make a drive that evening with the students, who will be taking their course finals over the next few days — it's the end of their term — after which their instructor, Jeff Munterig, will take them down to the coast for some R and R, though not before a celebratory party at Palmwag.

Earlier Round River students have compiled an impressive array of oral histories from Namibia, charting cultural attitudes toward rhinos, and have found that although there are occasional anecdotes here and there of someone's old uncle being gored by a rhino, and of men killing rhinos during the war to trade the horn for a bag of sugar, the rhinos were remarkably free of the deep negative cultural attitudes that would have presented an extra headwind to any developing community-based conservation plans.

The villagers don't have to love the rhinos in order for these plans to be effective; it's just important that they not *hate* them. (A new employee of SRT, Simpson Uri-Gop, has recently received a degree in conservation biology from the University of Kent, with his own research exploring regional and local attitudes; and as Mike pursues his doctorate, Simpson will be taking over more of the operation involved in running the Palmwag field camp.)

We follow an old rough trail along a sand river and up a narrow valley stippled with boskia trees, also called shepherd trees. During the heat of the day, goat herders often take refuge beneath such trees, in their lollipop-shaped lozenges of shade: the trees themselves white-stemmed with full rounded pastel domes. (As with the *Euphorbia,* the boskia trees are well spaced, though not quite as distant from one another as the isolated *Euphorbia.*)

The students have been up this canyon often, and went on a three-day hike into the dry mountains beyond, where they found several rhinos—the most they'd seen yet—and saw lions as well. This evening, we stop on a windy ridge and get out and admire the sweep of the plains below. Mike points out the tracks of an elephant—how amazing it is to consider such an immense animal living in such arid conditions: why not a giant orchid, or a jellyfish?—and, not too much farther on, he points to a giant, smooth-scalloped swath of sand dune beneath a mopane tree, where an elephant has bedded down out of the sun.

Such is the fineness of the sand grains, and the weight of the elephant, that what remains in the sand is the perfect wrinkled mold of an elephant: every crease of skin, every fold and wrinkle, is transcribed perfectly into the sand. It's a little like the snow-angel imprint that children make by lying down in snow and flapping their arms and legs—an elephant angel—except the sand is a far more specific medium, revealing the contours of the elephant's skin in such great and tactile detail.

It is a kind of super-illumination, or super-relief map of the elephant—a life-size monument of an elephant—and so fresh and undisturbed is the imprint that the effect is eerily and wonderfully that of looking at a kind of ghost, or some other time-lapsed phenomenon: as if the animal *was* here, and still *is* here, and yet is gone, too. As if it is gone but has left behind some remnant of itself so powerfully observed or felt as to achieve a presence after all.

This triumph of presence over absence cheers me, and the

effect is so striking that I find myself wanting to walk carefully around the sculpture, as I would were the animal still resting there.

It seems that we passed by this place only a few hours after the elephant got up and moved on—a few strides later, we find a sand-soaked piss puddle, still wet despite the day's heat—and across such a huge landscape, and across the tableau of seventy million years, is not the difference of a few hours then so insignificant as to be almost the same thing as instantaneous or contemporaneous?

Relatively speaking, are we not still looking at the elephant, smelling his scent, and breathing his just-breathed air?

We stare down into the valley, expecting to see the animal. The sun is setting, the long violet shadows are hurrying over the heated red rocks, and a bevy of giraffes—three, then four, then five—appear over the horizon, like walking skyscrapers, and descend the slope. I have to laugh out loud, so surprised and disoriented am I by the sight of animals larger than trees—animals towering *above* the trees, as if part of the world has been turned upside down; for if scale can suddenly be so upended, after forty-six-plus years spent with me learning it, then cannot all else, too?

And that such upheaval should possess beauty as its only side effect is a wondrous realization indeed, as unanticipated as the arrival of the giraffes themselves. I grin from ear to ear, then laugh again, feeling dizzily like a child.

The surprise of my own laughter reminds me of when my oldest daughter, then six, went snorkeling for the first time and peered down into a coral reef, in which darted and swam dozens of fish of all different bright colors and sizes. She kept her head down, staring, and began to laugh as she swam, the laughter sounding peal-ish and bell-like through the snorkel: a natural response to the sudden surprise of such industry, and such beauty, and the discovery of a magnificent world within a world.

How many times in this life do we get that chance, as adults, to see the world, or even any part of it, entirely anew?

The next day, Mike has arranged for us to go on an elephant-searching journey up the Hoanib River. It's our best chance to see elephants — daytime elephants, rather than the ghosts of tracks and football-size dung mounds — and because he is weighted down with paperwork, he has found a guide for us, Andreas, from the village of Sesfontein, not too terribly far from Palmwag. It's Andreas's day off — he works for SRT as a rhino tracker — and though Mike thinks that Dennis and I might be about to find our way up the Hoanib alone, he says he'd feel better if we had a guide. It's just so easy to get stuck in the sand, he says, and there's always the possibility of getting lost — there are no real roads, and certainly no directions — and then there are the lions, leopards, elephants, and the great et cetera: and certainly, we don't argue.

Driving out to pick Andreas up at his solitary little adobe house in the desert — he lives set back only a short distance from the road, but a great many miles from the community — we pass numerous springbok, gemsbok, and giraffes, all out strolling the basalt prairie in the cool of morning. In the presence of such bounty, I find that once again, as among the crystal fields, I feel compelled to count and gather, to acquire data, to measure and quantify, as if to authenticate or verify the wonder of what I am seeing. It is a pesky reaction, a response in part perhaps to a life-time of loss and going-away.

We are due to pick Andreas up at seven. The desert is still alive with movement, large herds passing back and forth from water or pasturage — and the little trees are brighter and greener after the night's cool, and the red rocks are cleaner and redder, and again, for perhaps the dozenth time already, I feel younger: not years younger, but a lifetime younger.

Dennis has commented on the physical corollary of this same feeling; as my spirit feels youthful once more, so too does his body. He was in a plane accident years ago up in northern Canada—he stepped off a floatplane in a fast-moving current, the plane spun around, and the still spinning propeller whacked his shoulder, nearly severing the arm. He had a satellite phone and was helicoptered out, fourteen hours to Seattle. It was touch and go for weeks, and even now he has but limited use of the arm, his left, and is in almost constant pain: not just the arm itself, but his back, neck, and head.

The heat of Africa, however—the stone-baked quality of it—is melting his muscles back into a looser, suppler thing. He says this is the first time he can remember being completely pain-free, and this even after the forty hours of travel it took to get here. All the thick scar tissue within is melted here, has become as flexible as taffy. And that's how I feel inside, thousands of miles away from the responsibilities and brittle worries of ceaseless activism on behalf of wilderness in the Kootenai National Forest in my home in the Yaak Valley. I feel as if, were I never to step back onto the battlefield ever again, things would be all right in the larger scheme and order; or if not all right, then little changed, or even unchanged, as the result of either my absence or my presence. And rather than that being a dispiriting realization or perception, as one might imagine—twenty-plus years of hammer and tongs so far, all for naught—it is instead strangely liberating.

I know that when I return to the States, I will be pulled back into the struggle—how easily our species enters into war!—but here in Namibia, with every reference and every experience new, my response to the world is not war but is instead peaceful; and the years of tensions within me fade and then vanish. They become so completely gone that it is easy to believe they will never return.

Off to our right, a single springbok is caught between the two flimsy parts of the rinderpest fence, the utterly imaginary before-and-after of disease/no disease. It races up and down the narrow alley, back and forth, seeking an exit, and not desiring any one side more than the other, but instead desiring only escape.

Andreas is ready when we pull up in front of his little pueblo. A thread of deep blue smoke rises from his chimney, and his wife accompanies him out to greet us. Andreas is a tall, lean man, perhaps thirty-five, or perhaps fifty-five—how can such things be in question? Is not twenty years a long time?—while his wife is much shorter. She is vibrant and buoyant, while Andreas appears to have had a hard night, though he is valiantly pleasant, quiet and restrained, and clearly hurting: holding himself carefully, as if under the grip of a splitting headache.

We ride in silence, and despite feeling so under the weather, Andreas leans forward and watches the passing landscape as if he has not seen it each day for those fifty-plus years, or however many there have been, but has instead been here only a week. His pained and bloodshot eyes pick out a cory bustard where we would not have seen it, and despite his extremely rudimentary English, and the travails of the morning, he endeavors to tell us how many eggs the bird lays, when the breeding season occurs. He points out ostrich, gemsbok, giraffe, and a lone kudu, the body type of the latter reminding me of North America's mule deer, though with curved and spiraling horns rather than dichotomous antlers.

We travel east, through the wild and rocky county that is just a little too arid for even the transient villages of goat herders to colonize; though as we descend from the basalt plateau country into the lower valley of the Hoanib, we begin to sense that man has been here, for all vegetation below a height of about four feet vanishes and the earth is covered with scalloped dunes of orange sand, ground to the finest possible consistency.

The impression is that a wave of fire has rolled through, melting the earth to sand and torching all life up to that six-foot height—a breath of nuclear fire melting stone to sand—and back in the trees, we begin to see the tents of goat camps.

We enter the village of Sesfontein where all the soil is a packed and powdery sand, the result of centuries of hoof-cut trampling, with the sharp eyes of the goats seeking out and nipping with young sharp teeth every possible sprig of vegetative life. *Sun* equals *plant* equals *goat meat* equals *sand* is the equation here, so that the sun now seems to be pouring down eternal columns of orange sand onto the land around Sesfontein, with the village and the goats and the landscape itself drowning in sand.

A funeral procession is proceeding through Sesfontein. Two donkeys pull the wagon, their little legs scissoring furiously even as the wagon itself travels at a crawl. Behind the wagon are numerous old cars and trucks, unearthed, it seems, for only this occasion, for it is more cars than we have seen on all our cumulative travels in the desert thus far—a good thirty or more vehicles, a veritable traffic jam, creeping behind the cart—and twice as many villagers walk alongside the procession, flanking the road and passing from sun to shadow; and by walking steadily, crisply, they are able to keep up with the motorcade.

The steadiness of pace and unity of direction conspires to give a sense of focused purpose to the occasion, though there is some-

thing about the scene also, and their pace and attitude and demeanor, that refuses to be categorized as urgent.

Beyond the cacophony of dress—some travelers are wearing old knit suits, despite the heat, while others are dressed in
brightly colored clothes that flash as they move from shadow to
sun, shadow to sun, like brilliant birds, and with still others wearing shorts and sandals—there seems an odd and undefinable and
saddening element of familiarity to the procession, a sense that
the travelers themselves are much familiar with this route, and
this practice.

They stride with their heads down, babies on their hips, none
of the travelers, not a one, possessing even a gram of fat: and although anyone can die anywhere at any time of any given cause,
there is something about the brilliance of the day and the quiet
and almost businesslike attitude of the procession that makes me
think of AIDS. No great sleuthing is required to make such an assumption—it's killing more than 125,000 people per year in Namibia alone (population roughly two million).

It could just as easily have been cholera, or dysentery, or mal-
nutrition. One thing it almost certainly was not was old age—
the life expectancy in Namibia for males is back down to fifty-
three, little different from many centuries ago—a solid twenty
years fewer than our own brief but shining years.

We follow the procession patiently, windows rolled down,
pale arms hanging out. Andreas says nothing from the back, is utterly silent. I am not impatient—on the contrary, I am pleased to
still be alive and in Africa, traveling with no timetable or agenda
beneath open blue skies and brilliant December sun—but I am
beginning to wonder where the procession is traveling, for we are
nearing the other side of town, the last of the shade trees, with
their goat-browsed dead space between ground and first branch.

The desert—limitless, brilliant, superheated—lies beyond, with no cemetery in sight, so that it occurs to me they could be traveling for another ten miles or more at this pace, in some form of ceremony.

In the last grove of trees, however—shrouded by fronds of feathery tamarisks—there is a dusty-windowed brick building, an abandoned-looking school, and it is into this dirt parking lot that the procession turns slowly—the pedestrians heated now, despite their broad straw hats and fans—and our truck leaves them then, travels on straight east.

We ride on, passing through folds of rock and sand, crossing dry riverbeds, and then come to our own unmarked intersection, and turn north. The road quickly turns to silky sand, where dunes have swept in over the road, and from the back seat, Andreas points out various alterations to our route—to the left of this tree, to the right of that one; as if reading the current in a rapid river, and with our truck trying always to bog down in sand. The road is long-ago gone—we have lost it in a matter of only minutes—and realizing that we are only moments away from getting stuck, Andreas hesitantly offers to drive, and we gratefully accept his suggestion.

The place where he is taking us is a region where the government came in a few years ago, along with some nonprofit conservation groups, and drilled two test bores along the Hoanib in order to provide a dependable, permanent supply of water so that the myriad wildlife inhabiting the region might survive in drought years, and so that mortality might be reduced by the animals' not having to stagger miles across the desert in search of another river, which might or might not exist within the range of even their most heroic efforts.

Hence, the elephants, living here full-time now: traveling up and down the dry sand river, browsing, and bouncing back and forth from one water hole to the other.

Is it right, or is it wrong, this engineering of nature — or is it somewhere in between? It seems, like too many others of our questions of the day, unanswerable.

The dunes grow larger on either side of us, but Andreas has found a seam, and we wind along that path and then crest a stony red skullcap of bare basalt and look down on the thick green clot of vegetation that is the Hoanib: the river seeming massive and major, so incongruous amid bare stone and bare sand, and with bare sand-colored cliffs looming above it.

We pause and look down on the curving jungle — glints of blue water braid in a thousand glinting strands — and we can tell even from here that the water is no more than Achilles'-heel-deep. I do not know which is more unusual and more beautiful, the blue shallow scrim of water or the rich green clot of strange low jungle: the most impenetrable maze and thicket of vegetation I have ever seen.

Even from the rise looking down on it, we can smell the sour-rot fecundity — the tangled riverbank such a living entity amid the all-else stone that it seems to us the jungle is only napping, and that at any moment it might stir and then rise, like the most gigantic creature yet.

Like a green factory, the land is producing an astonishing and magnificent stinky belch of green, a ceaseless exhalation — as green and overwhelming as the sun is hot — and we stare down at the dazzling marvel of it a while longer. After our having wound our way through the dunes, the river is simply not what we would have expected. It's so *contained* — less than seventy yards wide — but so utterly wild; almost otherworldly, until I realize, no, it is just not *my* world.

I imagine how cool and damp it is back in that matted green, how there might be low tunnels and warrens sneaking through it like veins and arteries, bringing even more life to the tangle.

"Are there leopards in there?" I ask, still staring at the ribbon of green, mesmerized.

"Yes," says Andreas.

Sticking up out of one of the dunes before us is an old milled plank and a discarded sheet of plywood with some lettering on it. As if exploring the ruins of some lost civilization, I lift it free of the sand and try to read the faded lettering, which is in English, but too many of the letters are already gone; I can make no sense of it.

"Elephant camp," Andreas says. He explains that it was an ill-fated tourist camp with tents and little outbuildings, where travelers could come and watch the elephants during the rainy season. But for whatever reasons, it just didn't work out. Maybe the people running it weren't the best businesspeople, or maybe the location was too far out in the middle of nowhere. Maybe there were other elephants elsewhere in Africa—tens of thousands of them, actually—in far easier locations to reach than this isolated little population.

The camp has only been out of business for a couple of years, Andreas says, and shrugs, pointing to a few other clues here and there, duned over or clotted over: a scrap of cloth, another piece of lumber; an old cot, a truck tire. Ghost barracks. The elephants came in after the people left, he says, and dismantled everything. It wasn't vengeful, he says, or else they would have done it while the people were still in their bunkers and barracks. It was more like housekeeping.

This tourist business, the guiding business: the kinks are not worked out of it, not in the least, and may never be, for if they were, it would not then be wild nature at which the tourists and travelers were gazing, but instead, a mirror—and who would pay to go look in a mirror?

We're fortunate to have Andreas with us. Not so long ago, he was guiding a documentary film crew and filmmaker into rhino country, traveling deep into the volcanic backcountry with camels, which are the preferred method of travel for many of the trackers. The camels are able to go places no vehicle can travel, and require far less care and maintenance. The camels are a little difficult at first, requiring a breaking-in period for each rider, in that upon meeting a stranger (much less one who climbs onto its back), the camel spends the first day reaching back and trying savagely to bite the rider, regardless of the camel's previous experience with being ridden.

It's just a kind of formality, a requisite code of behavior; and after that first day, the camel settles down and spends the next forty-eight hours only spitting at the rider—vomiting on the rider, actually, with the camel twisting its big neck around, in serpentine fashion, until the gigantic head is positioned to hurl the green putrescent vomitus directly onto the rider or, if the rider is dodging, at some part of him or her (they aim for the face).

This goes on constantly during that second and third day, but after the third day, if the intrepid rider has endured, the camel evidently comes to accommodate him or her; and the rider, likewise, has seen the worst of the camel, so that things now improve rapidly.

The filmmaker was a lady, Andreas says, who was dressed up in fancy clothes and shoes of some sort—fancy sandals, it seems he's describing—that weren't particularly well suited for tree climbing, which was what she and the rest of the crew found themselves having to do when the rhino they were filming charged them. In addition to her outfit, the lady's formidable physique conspired against her being a very good tree climber, so that even as the rest of the film crew were scattering like quail, scurrying to the tops of various boskia trees that fortunately happened to be nearby, Andreas was having to hold the affrighted camel string's lead rope with one hand while laboring to assist the filmmaker with her tree climbing, even as the nearsighted rhino was running amok, following any flashes or blurs it glimpsed and thrusting its daggered snout at any perception of movement, and at any scent of humanity.

The filmmaker kept falling out of the tree, and Andreas kept shoving her up there one-handed while sidestepping and dancing around the lunging rhino. Each time, the filmmaker would slip and fall on top of Andreas, but finally she was able to hold on, and perched like a partridge a scant six feet off the ground, quaking.

Andreas barely had time then to dodge the rhino's next jab, tucking in to the other side of the lead camel and using it as a shield. All of the other camels' saddles were slipping off in the stampede, and in their terror, the camels began breaking free of their halters and pack strings.

With Andreas clearly cornered and identified now — dancing around the camel, continuing to keep the camel between him and the rhino — the rhino simply thrust its horn under the camel and ripped upward, tossing it out of the way, leaving Andreas now completely exposed. (Miraculously, the camel would survive the goring.)

Forced to run, Andreas was barely able to make it to his own tree — the rhino's horn had come so close to him when it gored the camel that it had also torn his shirt — had he been a fat man, or had the horn been a few inches longer, he too might have been included in the goring — and from their various perches, the expedition waited for the sun-heated rhino to calm down and gallop off.

They sat in their branches a little longer after that, letting their own hearts settle, surveying the chaos of loose saddles and camera equipment, and the camels now as gone away as the rhino himself. It seemed unbelievable that no one was injured, and that even the gored camel's wound was, despite its profuse bleeding, a transverse surface wound—a swipe rather than a plunge—and after a while, Andreas and the film crew climbed down and began picking up all the loose pieces, and cinched up the camel's wound. Slowly, the runaway camels returned, over the course of the afternoon, and later that evening the film crew proceeded on.

Remarkably, the effect this fantastic story has on us is a positive one. It makes us feel that we are attended by great luck, and that all misfortune is behind us, and that the day ahead is bright and clear; even as we know and understand that surely the film crew must have felt the same way when setting out on their own blue-sky journey. I find myself remembering Mike's incredible photograph of the rhino charge out in Damaraland and wonder if maybe the odds of survival might be a little better than one would previously imagine, but then I realize the obvious filter, that it is only the survivors' stories that get told, after all, and therefore be always careful, always cautious; it seems again that I can scent the leopards lying coiled and heat-panting deep within that greater coil of green along the banks of the Hoanib.

We ease forward, passing through the ruins of the old elephant camp—the dunes are stippled with giant dung, and the giant prints of elephants—and on reaching the sun-struck alkaline shore (the thin sheet of water already fast on its way to vanishing, even less than a week after the end of the rains), Andreas turns onto the narrow path that follows the river downstream, as if entering a freeway on-ramp. Following the browsed swath of the elephants, he crisscrosses the broad glinting trickle of river, avoiding the quicksand and soft-slurry stretches, reading the braids of dune and gravel like a riverboat captain.

We find ourselves descending, choosing one slot canyon over another, with the sun high above us beating down into and bouncing off of the black-walled cliffs of basalt, and it seems utterly possible to me, perhaps even a mild likelihood, that had Dennis and I set off on our own, we would have gotten almost immediately stuck or lost or both, at which point we would have had to wait, or not wait, for some form of help, or mercy: stranded in the center of that green coil of leopards.

I don't mean to be fretting about such things like some *tourist,* despite my being exactly and absolutely that. I don't mean to slip so easily into the persona of those walleyed visitors who, on entering my valley in Montana, begin to tense and stare about wildly in search of the ten or twelve grizzlies we have left, about which they have been told the most lurid tales. But there is a little of that in me nonetheless, and it occurs to me yet again how much childlike joy and fear there can be in an encounter with the unfamiliar, how rare that is in all of our lives and, increasingly, in so many of our landscapes.

One slot canyon broadens back into a dazzling panorama of dune and sky—the river splays wide once more, so shallow that the glittering silver top-water minnows that skitter across it seem to be half flying, half swimming—and on the leeward side of one of the anchored dunes we spy a mud and stick hut, looking more like a little cage than any structure fit for habitation, looking like the skeleton of a thing, with the mud that had been packed between the teepee ribs cracking and falling away like rotten mortar, so that the bright blue sky is visible through them. The structure is barely three feet tall, and though it appears to us to be like something a sojourner might throw together in a pinch to escape a thunderstorm, or broiling heat, or bitter cold night, Andreas tells us that it belongs to a Himba traveler, a nomad, and that it is still in use, that such structures are actually permanent, used again and again by the same travelers year after year, being patched up only as needed.

And standing next to the little stack of dried mud and sticks —twigs, really—and thinking across the distance between my own life, my own dry, warm shelter, and this little rib cage of a pile out beyond the middle of nowhere—standing there right next to it, the structure barely more than an oversize dirt dauber's nest—I am not sure quite what I feel. Wonder, I suppose, and a deep unfamiliarity with the world.

This mud hut, mud home of the traveler—no more than a lattice birdcage, and barely the size of a small man sitting hunched on his haunches—seems like the sort of thing one might see in a travel book or a photo album, but it is not a photo album; it is here far removed from one world and yet in the center of another world. The heat is baking our pale skin, and the world is poised, paused, and yet the world is also rushing past.

The water disappears—the river has already gone underground —and now there is nothing but sand snaking in a tunnel through the trees, tall, river-nourished mopanes offering shade down in the canyon, and the winding scrim of trees a river of green within its towering walls.

Oryx are galloping in front of us, long tails floating behind them, rocking gait steady, muscles bulging as they power through the soft sand, across the dry river, and up through the dunes and back into game trails that disappear into the curtain of green.

In addition to the curtain of green, there are dead-standing spars and snags of trees that grew too stressed; for them, there are no boreholes coming to their rescue.

I remember what Mike told us: that in such a water-limited ecosystem, nearly all species are fugitive, with the requisite ability to get up and move great distances in search of water, and that one of the potential drawbacks to the establishment of permanent boreholes is that over time it will concentrate once transitory populations in a fixed place, and will attract still others, until sustainability is exceeded and the tipping point is crossed.

This happens all the time—this wobble, collapse, and rebirth —but one of the less publicized or considered concerns too is that in approaching that traditional collapse, the burgeoning populations will do such harm to the underpinnings of the ecosystem on their journey toward that collapse that the system becomes damaged beyond its usual ability to provide meaningful resurrections or rebirth.

In certain places—the ghost eddies of where the river once flowed, and where it will always flow again—heavier cobbles have settled out, so that an inlay of water-polished dark stones lies packed beneath us, and for a few moments it is like driving on the brick streets of Paris, with sun and shadow stippling us—and then the ghost river's signature changes and we are deep once more in the silken whisper of sand, and we feel immediately the greater reflected heat.

In other places—the shallower places, closer to the ghost shore—fantastic polygons of dried mud cracks shelve and overlap one another like scales of shed reptilian armor, or like a surreal collection of china and dinnerware, the curled plates and cups still darkened from last week's rainy season.

Elsewhere, crystalline salt patches sprawl gleaming like buckets of spilled paint, around which butterflies congregate by the dozens, feasting, gathering the minerals and nutrition that will help empower them to go back up onto the desert floor above, to aid in the pollination of so many of the desert plants and flowers, upon some of which the rhinos undoubtedly graze, so that from a far enough perspective, I suppose that a viewer could say that a butterfly *is* a rhino, or that a blossom is a rhino; and that a flooding, charging river of tumbling, clacking boulders and cobbles, drying weeks later to the sheen of hard-baked saltpan, is a rhino. In this made and healthy world, it may be that there is *nothing* that is not a rhino, and yet, we almost ran out of rhinos, and one does not need to be an astute student of history to recognize that Africa is tipping, and—back to Pangaea—the world itself is tipping.

Somehow, though, down on the Hoanib, it seems less frightening to consider such things, mining with our eyes this deep and rich vein of life. We ask Andreas what if anything lives up above us, up at the tops of those cliffs, and he says there's nothing up there but baboons and leopards, that the leopards go up there to sometimes catch the updrafts, the cooling breaths and breezes of the canyon—the respirations—and the baboons perhaps likewise. But that there's nothing else up there.

He doesn't say it, but I cannot help but think it: that the Eden is down here, right now.

And up above, beyond the leopards and baboons that perhaps sit perched at the lip of the canyon, soaking in those breezes, panting, back in the shadows of one cliff or another—beyond that, out onto the planar spill of the horizon, and in another world, a different level from this one, the armor-clad rhinos wander, isolated. Miracles, passing back and forth, for tens of millions of years, and with anyone, almost anyone, able at any time to climb up out of that canyon, that Eden, and peer out across the desert wilderness and to behold other miracles beyond our own.

The things that never occur to you, in this world: the rising heat of the sun makes the sand softer, so the driving becomes more difficult. The sound of driving across the polygons of mud cracks is exactly like that, I suppose, of the breaking of a thousand clay pots—a releasing disorder, and not an unpleasant sound; and what is in us to make this be so?

We've been seeing elephant tracks for so long, and in so many places, that it's almost as if we've stopped thinking about the animals that made the tracks and have become accustomed instead to perceiving the animal only through tracks; as if the imagination is, through familiarity, draining away.

Is this how it is with extinction, I wonder—where the whole, no longer known, dwindles to longer and longer absences, with the species finally blinking out so quietly that its last passing is anticlimactic; that in actuality, the extinction began and evolved long before the physical fact of the animal ceased to exist?

So it is that we barely recognize the animals standing motionless in the trees before us: hulking silhouettes in the shady grove of an island in the middle of the river of bright burning sand. Andreas points them out—*Elephants,* he says, and a part of me wants to correct him, or my own eyes, and say, *No, mountains.*

The elephants, or silhouettes of elephants, are all staring, watching us, and Andreas stops the jeep at a respectful distance to let them become accustomed to our rude intrusion. They remain motionless, though something about the unity and stolidity suggests that there is a communication going on between them, in that perfect immobility; and finally, as if a consensus has been reached, the broad flap of one ear stirs, fans the stillness of the heated shade air in which they are all standing shoulder to shoulder, and the muscle of them shifts and seems to relax a notch or three; though still, defensibility and unity remains the potent message projected.

Andreas drives forward carefully, easing the jeep over to the far side of the riverbank, sand and cobbles scrunching under our wheels, with the elephants shifting to watch us.

We stop again when we draw even with the elephants, about fifty yards distant, so that Dennis can take pictures. Andreas is jumpy, utterly tense — just this side of frightened, it appears — with every aspect of his demeanor radiating distrust. He still has the clutch depressed with one foot and is keeping his right foot on the accelerator, ready to pop the clutch and bolt. And the fact that the man who has been almost gored by a rhino is far more frightened of elephants is not lost on me.

Still, there is something almost overwhelmingly attractive about them: something that makes a person want to trust them, know them, admire them. There is something about them that we are drawn toward, something we see in them that I think we want to see or pretend is in ourselves — a presumed camaraderie, so that we are tempted to assume what I suspect feels to the elephants like a forced bonhomie.

We are awed by their obvious power and strength, and by their intelligence, and by the presence of their more tender emotions, even within animals so strong and fierce — and yet I think there is a part of us that empathizes with what we perceive, mistakenly, as the physical clumsiness or psychic isolation of elephants. We perceive that they have baggy wrinkled skin, and that they shuffle with heads lowered, trunks hoovering the ground for scent, and we think, *Here is a fellow traveler for whom the world is still a riddle,* and we are tempted to seek them out.

We observe how there is no other animal in the world even remotely like them, and we misperceive that in that solitude there is loneliness. We move toward them with patronizing gestures, offering them positions in our circuses, and in our labor camps, and in our zoos, and are surprised somehow when things do not work out according to our plans. We shoot them and saw their tusks off to make into jewelry, in an attempt to make our own selves appear more beautiful and less shambling, and then are surprised and disappointed when they "turn on us."

Some of us chop their heads off and stuff them and hang them on the walls of boardrooms, hotels, and hunting lodges, and then are disappointed, chagrined, when they in turn reciprocate, attempting, sometimes, to grind us back down into the dust from which the fevered dream, the tortured or at least confused idea of us, first arose.

Across that distance, we behold each other, here on the Hoanib. It could be said that they are able to exist here now year-round because of our benevolence, but it cannot be said that they are our guests; on the contrary, it is they who are our hosts.

With the elephants as well as ourselves repositioned, we can see now that there is a baby among them, an *infant,* by comparison to all the others around him—and that he is heart-meltingly cute. He, alone among them, possesses no reserve, and appears to be trying to scamper out of the herd and come over to where we are, to investigate, or even play; but the adults, the old aunts and uncles, matriarch and patriarch and older brother and sisters, have him surrounded in a corral made by their thick legs, and every time he tries to scooch through, one of them pushes him back gently and shifts a leg in such a manner as to refortify the prison, the containment, the safe house constructed of the herd's legs; and finally, exasperated, the baby elephant contents himself or herself with peering out from between two of those gigantic legs, its head squeezed between the column of them but nothing else, with its little trunk twisting and twirling, trying to take in our curious scent.

Dennis and I are delighted, and feel that we could just sit there and watch the elephants watching us for hours—we grin at Andreas, and he smiles back, though weakly, trying not to dampen the experience for us with his adult knowledge of guide responsibilities, and his knowledge of elephants.

One of the elephants snorkels up a snootful of sand, then lifts his trunk over his head and sprays himself with a shower of sand, hosing himself down with it, and when we ask why he's doing that, Andreas says that it's one of the ways they cool themselves in the heat: the cooler, shaded sand bathing them, as do the tiny breezes created by the sand sliding down the slope of the mountains of their bodies, and with the sand perhaps even absorbing some of the radiant heat and carrying it away.

Frustrated, the baby elephant begins making odd little chirping sounds and fans his ears energetically, but the bars of his prison squeeze together more tightly, and with an audible sigh he plops down to nap, lying down in the sand like a man on a couch. Andreas eases out on the clutch, tests the forward motion of the jeep to make sure everything is still working—that we haven't somehow become bogged down in the sun-softened sand—and feels compelled to tell us finally that a friend of his was guiding some tourists on a similar elephant-watching journey, on which everything was going fine until the guide discovered that their Land Cruiser had gotten stuck, and when the elephants likewise came to understand that the truck was stuck, they charged, as if thinking, *All right, fish in a barrel.*

The guide had a gun and was barely able to turn the elephants by firing at them, striking the lead female in the ear, drilling a hole in it that did no more damage than if one were to fire a shot through a palm leaf.

Dennis and I listen to this and then look back at this herd of elephants, trying to discern any such capacity for betrayal, or what we would perceive as betrayal, but can find none; all seems to be only indolence, utter sun-struck lassitude.

Through Dennis's binoculars, we study the intricate striations of thick skin, with those myriad wrinklings (which increase the surface area exponentially, allowing even greater mass to aid in the dumping of excess body heat) matching exactly the striations and planes and laminas of cross bedding in the sandstone bluffs behind them, the cross section of sedimentary time resting beneath the caprock of basalt above: as if the elephants have literally arisen from the sand, stepped out of it and into the land of the living still possessing the mark of their origin and the stamp or imprint of their maker, and the desert's maker.

Lying in tangled flood piles on the braided river's island are the flood-polished spars of the trunks and limbs of acacia and mopane, barkless now and as smooth and brilliant, in that sunlight, as bone itself, or ivory. Indeed, the driftwood piles of spars that are resting next to the resting elephants appear identical to the tusks of the elephants—appear to be constructed of their ivory tusks—and yet we pay no mind to the driftwood spars, desire only the ivory tusks, and I do not, cannot, understand why.

Did some opportunity for choice exist within our own minds, our own path, not so long ago—six of one, half a dozen of another—wherein we chose an attraction to one over the other?

And if so, *why*, and how might things have been different if we had somehow chosen an attraction to beauty that did not involve killing but that instead would have been content, or more than content, preferring a selection of beauty that avoided killing?

It seems these days that one might as well be imagining a different species, and a different outcome to things, if not a different world.

We pass on by the elephants, not wanting to stress or disrupt them any further, and take our lunch a few miles farther on, sitting on more of those curved and gleaming riverwood spars. Here in the shaded corridor of the Hoanib, birds are still singing, even at midday, and even with a heated wind rushing down the canyon. It is as if a curtain is rising—so much life is stirring, and to me all of it is new—and yet there is also a part of me, the part that reads the news every day, that cannot shake the feeling that already the play is half over. Not just that section of the play or pageant or spectacle in which we have a part, but some longer-running performance.

Is this atmosphere of leave-taking all in my mind, or is it out on the landscape? Surely it is contained within me. How could billion-year-old rocks, or the ancient rhinos and ancient elephants, be projecting anything other than continued immortality, with the first billion years of life as insignificant against the scale of all-yet-to-come—as but a summer's day?

Somewhere in the acacia trees above us, a locust is buzzing its shrill heated summertime call—this and only this is familiar to me in Africa, the same sound as that of my childhood in Texas, and of my time as a young man in Mississippi—and I realize that it is a sound we have been hearing all morning and all afternoon, the unbroken sound of heat: as if either the same locust has been with us all this way, and all this time, or an unbroken chain of them exists, like strands of telephone wire, an unseen current of electricity sizzling in the sun, one of an infinite overlay of such currents, each one unseen but nonetheless part of the larger braid that holds the world together.

How many of those strands, those fibers, can dissolve, or lose their electricity, their vitality, before the structure or logic or mystery of this world begins to fall apart like something unwrapped? Nobody knows. But I think almost any of us can sense that more strands are vanishing each day. Some are thicker than others— some are even visible to our bare eyes—though sometimes it is the idea of the minute ones, the invisible ones, going away—the never-known ones—that frightens me most.

We spy another elephant up ahead, an hour later: an old bull of such size, half again as large as any of the others we saw previously, as to make the others look svelte. The others had appeared magnificent, but this old fellow looks as if he could change the world—as if, in fact, in his distant younger days, he might have had a hand in such activities: shoving aside huge masses of earth, changing the flow and gravity of things, rearranging constellations, damming old rivers, creating new ones, and fashioning new life out of dust and clay.

I don't know why this elephant isn't with the others. I don't know anything! How many days, how many years, how many lifetimes would it take to learn even a trace about this country?

The old elephant beholds us, and unlike the others, he displays no suspension of judgment, no wait-and-see immobility; from the very beginning, he evinces a hostility, if not an aggression, that is smoldering. Andreas pauses, knowing that we will want photos, but looks stricken, and gestures toward the elephant's ears, both of which are flapping slowly, steadily, like the wings of the largest butterfly in the world. One of the ears has a neatly drilled bullet hole in it—perhaps from yesterday, or perhaps from half a century ago, we don't know.

Six gemsbok surround the elephants like courtiers, ant-like—Andreas has told us that the oryx follow the elephants from one acacia to the next, to eat the seeds that the elephants knock from the tree while browsing (a high wind will also stir the gemsbok into foraging, traveling from one tree to the next, for this same purpose, so that again I have the heat-struck thought that the elephant *is* a kind of wind, made visible)—and the elephant departs the gemsbok then and begins advancing on us, walking carefully, deliberately, but with his ears still flapping, and Andreas lets out on the clutch, eases forward that same distance, then stops again, while Dennis's camera clicks.

As if hypnotizing us, the elephant continues to approach us, and we recognize that even though he's trying to appear casual, he is still traveling in a direct line toward us, and that despite the appearance of *ambling,* the distance is closing. Andreas grows even more agitated as Dennis fiddles with his camera. And watching through the binoculars, I can see that now the elephant is doing something funny with his feet: not ha-ha funny, but disturbing funny.

Although it seems impossible for this to be so, the elephant appears now to be trying to shrink, almost crouch, and his feet are coming down softer and sneakier, like those of a cat approaching a thicket of grass in which a mouse or vole is hiding. It's ridiculous: this six-ton beast is out in the broad middle of the day, out in the middle of the sand river, bigger than life—does he think we don't see him?—and yet the secret of his blood is betrayed. He *does* have damage on his mind. As he prepares to close that final distance, his body cannot help but obey the old habits and instincts of countless other stalks. The elephant knows there is not tall grass between him and his quarry this time, but the other parts of his body that are wired to both his memory and his intent do not care.

It's just a little thing—this crouching and sneak-footing—but the elephant is much closer now, almost close enough that should he choose to charge quickly now, he might be upon us before we can act. But he will not find Andreas napping, and we are already lurching forward. The elephant stops his approach as soon as we drive off; and for long moments afterward, the linings of our hearts tingle and sizzle with the delightful electrified cleanliness of survival.

Giraffes, ostriches, and baboons stride and trot and scurry across the dry riverbed before us, coming and going at cross-purposes, with the scene looking for all the world like throngs of pedestrians crossing a crowded city-street intersection. It is of course the borehole that has concentrated them in such density. They remind me of patrons at a bar awaiting happy hour. We have reached the end of the Hoanib, or rather, the last borehole; from here on out, it is just heat and sand and light, and the river itself disappears farther underground, so that even the bank-side vegetation thins and then vanishes, as does the multitude of life that we've been witnessing thus far. We turn around and head back, passing many of the same lingering individuals—the old sneaky-footed elephant is gone, fortunately—and then Andreas takes a detour, up and out of the cobbled river bottom onto a trail he knows, a shortcut.

We travel for hours, bouncing and jarring in the heat. Dennis and I nap, sun-lulled, despite the bouncing, and as the hour

grows later, Andreas drives harder and harder, until he is redlining the tachometer up steep graveled slopes and power-sliding on
the downhill curves, threatening to pitch us up onto two wheels,
threatening to tilt us over: and still, despite the beating, Dennis
naps on, sun-struck. In the back, the refrigerator rattles and clat-
ters, the fragile compressor vibrating like a jackhammer.

I cannot imagine what demons might be circulating in Andreas's head—does it have anything to do with the hour, which
is now approaching five o'clock? On he roars, zooming in a red-
dust funnel across the desert landscape, with the Hoanib now
somewhere far below us, a secret seam or vein. How much more
hopeful it would be to think of it as some wonderful middle-
earth not yet born, rather than one that has already been born,
but that might now be going away, due in part to our own appetites, and our own clumsiness.

It seems there is nothing in the world now that is not about
oil. It might appear that all stories in Africa are about water, or
the absence of water, and to a large extent they are: the heat and
solar radiation and aridity have shaped and sculpted almost all of
the ancient stories. But the burgeoning simmer, the now lightly
bubbling approach of global warming—the lid on the pot of
simmering water beginning to rattle tinnily—is intruding, creating new and often precipitous tipping points for every species
still in existence; and the grand and the glorious, and the highly
specialized, are out on the edge, in these new conditions, and it's
hard to know exactly what to do or where to do it.

Maybe the boreholes will help get them through this time of
hardship, or maybe they will aid and hasten the unraveling of a
finely tuned system, tested keenly against the force of time. We
just don't know.

Best of all, of course—for rhinos and giraffes, for grizzlies and salamanders, and for human beings—would be to stop using gasoline, and to develop a clean, safe, alternative form of energy available worldwide, before we descend too far into our own wobble, over on the back side of our own tipping point.

On Andreas thunders, stony-faced, nearly impassive—only the faintest display of consternation crosses his face—and as the country flashes past us, I see that we are crossing a landscape now composed almost exclusively of fractured, glittering quartz crystals—a hilltop as brilliant as a glacier—and it feels more than ever as if we are on some other planet.

Clearly, Andreas is late to some place, we have kept him too long, lingering with our silly tourist photographs and our African raptures—but by communicating my discomfort, I am able to encourage him to slow down a bit, for safety's sake: though still he is white-knuckling the truck along, not so much driving it as battering it.

We blast through the expanse of quartz fields and descend back into the red desert. The country begins to take on that look of nuclear annihilation once more, so that I know we must be getting near the land of goats again: and sure enough, after only another mile or so, a wikiup comes into view, and then a corral, and a couple of pueblo structures.

From out of one of these structures a young boy comes running, having apparently only heard us just now—I can only imagine how infrequent must be the passage of any vehicle, in this country—and the young boy, an adolescent, is running with his arms out, trying to reach the road before Andreas passes—we can see him coming from a long way off—but he does not quite make it, and Andreas roars on past, never looking back, red dust pluming.

It seems almost as if he is trying to outrun the day.

We encounter another denuded village, the sand dunes made as fine-grained as powder from an infinity of hooved tramplings, and the sand bright red in day's waning light, with a single store constructed of cinder blocks, and with but one tiny dusty window, adorned with thick iron bars. Andreas skids to a stop, and as if in a dream, we go inside, where a woman is standing behind the counter. There is nothing on our side of the counter, and on her side, there is little more. The only food in the entire store is stacked on a single high bookshelf behind her—it's all canned milk or sardines or chili—and I'll bet there aren't more than a hundred cans total in the whole store.

Andreas stocks up, buying this and that—having no list, but eyes roving hungrily, adding on from time to time—and Dennis offers to purchase the food in addition to paying a guiding fee, which Andreas accepts gratefully, and his agitation seems to be checked somewhat, though still there is a tension in him, not quite an anger, but a hardness, a sorrow, a despair, and a captivity. A flailing, a whirling inside, it seems to me, though maybe I am only imagining all of it; maybe the disorder is only within me, and it is I who am the disorder on the landscape.

We drive on, more slowly now: easing back into the fever of the reddening sunset. Passing one home, we see two small children playing next to a table that is set up beneath a small grove of shade trees. One of the children is standing and the other is sitting—they are playing some game with fistfuls of sand, building castles, perhaps—and they appear a little lost-looking. My eye shifts from them to the table, and I see that curled up on the table is the gauntest of women, wrapped, despite the day's heat, in a purple and gold blanket, and she is stiller than still. A column of dust-stirred gold light surrounds her, and such is the pathos of the scene—the weight of it—that it seems possible to believe the children have brought her out to that spot and placed her there in that gold light as if for some sort of remediation, or even cure.

It seems likely that she has been in that position all day, has not moved a muscle all day—the wind stirs her blanket, somehow making her seem even more motionless—and then we pass on. A sand fox skitters across the road, seeming to float, to fly, then is gone.

We reach Andreas's home in the last available red light—the tiny pueblo appears to be inhabiting the last pooling of tinged light—and it is with some final relief by Andreas that he parts from us. We all climb out and shake hands, and then watch as he hurries into that last darkness, carrying his double-arm bagfuls of groceries. His wife has come out into the front yard to greet him—was she worried, have we stayed late?—and it would be a tableau of archetype, the yellow-lit squares of lantern-light awaiting the homeward-bound traveler at long day's end, the glow of hearth tiny but solid, surrounded by the immensity of prairie, except that is no glow or lantern-light, neither candle nor flashlight, only that last going-away dimness of the sunset.

We drive on to our own tiny habitations out on the stony prairie, beyond the edge of the edge, but not quite into the rhinos' territory, the no man's land right on the border of where they begin to be seen again—the great knife-cut visages of Mike's favorite mountains, unnamed, are silhouetted by stars. In our encampment, we see a few quiet lights, no buzz or grumble of generator, only leftover solar muscle from the day's sunlight—and Mike and Jeff and the students are still gathered at the campfire, finishing their supper, and are eager to hear about what we have seen: the inventory, the circumstances.

Around the coals of the cooking fire, we visit about their goals. One student—an aspiring rap musician—has grand dreams of using the entertainment industry to mobilize public sentiment toward protection of wild country, explaining that music is the only medium that can reach or address or move young people. He talks about envisioning a generation of musicians and entertainers who are nowhere near as self-indulgent as generations of past and current performers—of a fierce and almost frantic ethos arising in which social and environmental activism is neither hobby nor duty nor occasional benefit, but a lifelong commitment.

"It's got to become *hip*," he explains, "or we're fucked." And there are none of us who would disagree with him, and we wish him well, and wish him endurance, too.

The students drift off to their dorm to study, and to sleep. Mike and Jeff and Dennis and I sit up by the fire, listening for lions, and visiting. Jeff is working on a tiger conservation project in China. When I was a young man, everyone thought Russia was going to rule and change the world. How could anyone be so wrong, and so nearsighted, over such a short span of time? Of course we had heard rumbling and whisperings about the *what if* of China—but it was uncomfortable to consider, and so we avoided it. The energy crisis of the 1970s, likewise. The trough ran low for a summer—not dry, just low—and we got a little scared. But then it filled back up and we went right back to it, snuffling and snorkeling, eyes shut. Thirty years rushed by.

Jeff's been administering the final exams all day, and Mike has likewise been pushing papers, trying to clear or at least reduce somewhat the volume of correspondence from his desk in order to go out searching for rhinos in the field with us in the morning. Dennis and I were the ones who got to go out into the

world today, and who saw the unending scroll of fantastic creatures all along the Hoanib, and we visit about that, recount that a
bit more, before all retiring to bed.

So passed one day of each of our lives.

In the morning, the nifty little refrigerator is broken, doubtless
the result of Andreas's mad cannon-ride run down off the red
mountain, so Dennis and I are forced to eat steak and eggs for
breakfast, and pork chops, too. We barbecue, grill, and pan-fry a
dozen eggs, six little pork chops, and two big T-bones, and then
eat them before they spoil, saving for lunch that which we cannot finish, but still feeling every inch the gluttonous Americans.

We finish with a cup of hot coffee in the already hot first light
of day, and then gather our daypacks to go out into the heat to
look for a rhino. It is overdramatic to say that I have lived forty-
six years in search of one, but by the same token it is inaccurate
on the low end of things to merely say that I'm forty-six and
have never seen a rhino in the wild. The truth is somewhere in
between, and as we prepare to go out and meet it, I'm a little bit
anxious and excited: deliciously so. Having previously believed
myself to occasionally have a somewhat robust imagination, I find
myself now facing the possibility of something I have no ability
to imagine — no preconception, no expectation, no reference —
and the prospect is intoxicating, cleansing, liberating.

About our chances, Mike has said that they are good.

We strike out into the heat, riding standing up in the back
of the truck with Mike, gripping the iron rails of the rack that's
bolted upright for that purpose, rhino watching. We swivel our
heads in all directions, like ship's mates high in the crow's-nest,
searching the horizon. The red desert, stippled with farther
mountains to the north; the black distant reef of the Brandberg

Massif to the south; more red desert back to the east; and more desert to the west as well, stretching down toward the coast, where, almost beyond the range of our vision, a thread of luminous silver marks the Skeleton Coast, the place where the sand and heat finally runs out and where the South Atlantic currents meet the shimmering heat of the dunes to give birth to that fog each morning: turgid cloudbanks looking for all the world like rain to the west, rain-a-coming, but never moving: vaporizing, vanishing, by midday.

"It's cold out there," Mike says, pointing to the little dream of clouds that hug the western skyline at the intersection of that curve of earth. He shivers. "Maybe the coldest place on earth. I've never been colder than on that coast. You never get warm there," he says.

The heat penetrates, chases all images of chill from us, and we stand there a moment longer, soaking it in, as if to store up some reservoir of its warmth for possible release later, should we ever venture into such chill.

"Nothing here," Mike calls out, and the drivers—Himba, Leslye, and Joseph—move on, with us gripping the rails once more, the truck bouncing and jouncing and squeaking over cobble and stone, and with the riders in back surveying the terrain both far away and near.

In every sand wash we find elephant tracks, but they are old and likely do not reflect any residential animals but instead passers-through traveling from one distant spring to the next.

Finally Mike and the drivers spy some rhino tracks—a lone male, it seems, from the scrapes and pawings—that travel straight down the road for a short distance before veering back into the desert, the sun-varnished black-red vitreous sheen of each cobble glinting in morning sun like the innards of some just disemboweled fowl, and an infinitude of such cobbles spilled across the desert. How can any animal travel any distance through such a landscape, much less *run*, as do most all of the desert species, on frequent occasion, while either chasing or being chased?

The tracks are old, possibly as old as twenty-four hours, but they're all we've seen, and we're eager to get out of the truck and walk, and so we do. The trackers say that there is a spring not too far ahead, and that it's likely that's where this animal was headed, although he probably watered there and then passed on, traveling on to the next spring and the next—traversing the desert in nearly impossible transects between points of moisture.

We set out walking, and it feels wonderful, learning the country with our feet, and with the muscles in our legs and backs. It feels like opening a great novel to page one, with the first hint or promise of the unimaginable richness lying within and beyond.

Watching the trackers work, I am surprised how wandering the rhinos' travels are. Any feedlot animal would merely track in a straight line, and would stick to the graded road, as well—as would almost any human. But for this animal, the spring—even amid such heat and aridity, and with the three thousand pounds of him surely operating on at least a low simmer—seems secondary; his real calling, it appears, remains the *Euphorbia* bushes, and other unknowable points out on the farther landscape: other odors, other secrets.

As if following the pheromone scent of bees in clover, or the intricate circuitry of ant-dance, Himba and Joseph and Leslye fan out and stroll through the desert with their heads tipped down, often with their hands clasped behind their backs, studious inspectors attending some other zone: as if summoned by other music, or guided by other directives, other rules of logic, and other clues and evidences of what was once here but has since moved on. And again I am struck by the paradoxes between nearness and distance: the thin veil or shadow between here and not here, against which we and the trackers seem to labor. After seventy million years, what difference and distance can there be, really, in a twenty-four-hour absence? Against such scale, is not the rhino just as good as standing before us, huffing and puffing, and utterly alive in the moment?

And yet there is nothing; or rather, we see nothing.

Like birds wading the shallows with their heads down, intent on spying a shellfish, a clam, a scallop, the trackers stalk the desert, betranced by the gone-away conductor; and like birds, they sometimes wheel as a flock when the conductor instructs them to turn left or right, though other times the current of them diffracts, spools out into the desert in a wandering triumvirate, as if lost briefly, or backtracking, so that again, particularly if seen from far above, their movements take on the character of a swarm of ants whose nest has been disturbed.

Like everything in Africa, the perspective is tilted a few degrees—sometimes radically different, other times only a little so. The various directions of the trackers, when they unspool, trying to parse the script where it has disappeared onto the bare stone, are like those of disturbed ants, and yet even down at ground level and in real time, the pace is such that it might seem the trackers are being viewed at a great distance.

It's a tiny thing, this difference or disparity. But it is a tiny thing, like mortar, between so many other larger things; and we follow the trackers, who, despite their studied attention to the barren ground, and despite their occasional wanderings and backtrackings, are moving much faster and more gracefully than Dennis or I.

The difference is such that sometimes Dennis and I feel compelled to break into a trot, which only makes things worse; we stumble even more, the rocks like spilled marbles underfoot, each rock the size of a clenched fist. We stub our toes and slip, lose our footing, trip as if in pratfalls.

How does someone fall down while walking on perfectly level ground?

The trackers continue on, pulling away from us as if working in some strange seam between space and time, drifting as slowly as continents, yet with a significant distance opening up, over what seems like a short time, between us and them: and now and again Dennis and I glance quickly at the ground, as if believing that by miracle or providence alone the clarity of a single track might manifest itself to us.

Ahead of us, the trackers widen and gather, disperse and coalesce, following that which we cannot see. Is it all a grand joke? Are we being had?

Like whirligig beetles in a pond, Mike, Joseph, Leslye, and Himba pursue that which might as well be only a dream. It is a dream they have seen often—a dream for which Leslye, Joseph, and Himba are rewarded financially, whenever they can attain and document it, and for which Mike is rewarded—how? is it fair to say spiritually? I think it is—each time he attains it.

There is no getting around it, watching him: how much between two worlds he seems. He is the field director of SRT, and the trackers work for him, and yet he is among them, is one of them. He is white and from a privileged background and they are black and from strife and war and famine, and yet there is also no difference, there in the shadow or track of the rhino. There is no difference, and they each follow it, and we follow them.

Will he ever go back to England? It's hard to imagine.

Will he spend the rest of his life following rhinos and living in solitude?

No one can imagine accurately the future. We can only follow.

We come to the spring where we had hoped the rhino might be napping — having ceased, atypically, his peripatetic travels to chill for a day or longer, as if only waiting for us to catch up — but there is nothing: no rhino.

The spring is delineated by an oval of pale green salt sedges, thickly clotted and with surprisingly little other vegetation, as if among all the plant species in the world, only these sedges can survive here; but here, they prosper.

Not even the rhinos eat these sedges. They grow tall and dense, hydra-headed yet symmetrical in their radial splay, each plant a cluster of sedge-antennae taller than a man — certainly tall enough to conceal a lion in their thicket, or a rhino, or anything else — and as we proceed down the little warren that snakes through their midst — the narrow trail to the water — I am acutely conscious that we have not a single weapon among us. Even a spear would provide some reassurance, but we have nothing: only the flat teeth of herbivores, our canines rounded by age. The closest thing I have to a weapon with which to defend myself is a ballpoint pen.

No one's even wearing a knife. It's a singular feeling, proceeding on into that maze of sedge, knowing there could be a lion hiding in the sedges, taking refuge from the rising heat — napping, perhaps, only to be roused irritably by the downwind reek and noise of our approach, or, far worse, already awake and lying in wait for the next traveler to come to the water, as all must.

We have barely entered the thicket when we spy an immense and incredibly fresh lion track.

I've seen dozens of mountain lions in Montana, have had them stalk me in all seasons—a disconcerting event, always, whether I'm with or without a weapon—but the unease I feel—no, it's not unease, it's true-blue fear—is off-scale compared to the heart-hammering fear of the unknown, flavored with just enough extrapolation of the *known:* the freshness of the track and the size of the track, easily twice or three times the size of any lion I've ever seen in Montana.

Mike and the trackers say that the print is probably from last night, and seem to evince no fear or even caution, and I wonder, *Who am I to come into their country and tell them what to be frightened of?*

But the farther we travel into the sedges, the tenser I become, until finally we come to the water, shallow and crystal clear, pooled in a bestilled fountain with a black and gray marl as its base, and with the shoreline stippled and trampled and pocked with the tracks of every hoof and claw, it seems, present in this desert.

The sedges are raspy, like the teeth of saw blades—there can be no shortcutting here, no deviation from the one trail woven through the knotted mass of them. Nothing could penetrate the sedges on either side, and their all-one-ness, with the absence of any other plant species, is puzzling to me, for it is my understanding that in most harsh or extreme environments, numerous species adapt in likewise extreme ways, being sculpted by the land in fantastic ways to fit the narrowest of niches.

And the salt sedges *are* utterly fantastic. But where are the other experiments, the other baroque ornamentations? There seem to be none, only silence and absence—only sedges drinking and filtering the spring's lucid waters, which Mike says is clean enough for us to drink.

It occurs to me again that whoever or whatever designed this world, here in Namibia, and in the beginning, got it right the first time out, and that some of the slightest variations since that time, while perhaps not able to be classified as right or wrong, have proven to be more enduring than others: though we should not fall into the trap of confusing relative permanence or length of story on earth with "success."

I have to admit, as a geologist and biologist, I fall easily into the allure of time, and am doubtless overly impressed by the admittedly abstract quantifications and measurements.

And yet time, the animal or organic phenomenon of time, is rife and radiant with not just the abstract, but the specific: it comprises all the senses, or is at least structured in such ways for the various senses to attach themselves, like chemicals or molecules: like organic compounds bonding to a larger organic compound. Time is both particle and wave; surely it tilts, cants, stretches, and compresses like the compounds, and even the continents themselves. Surely time separates, then comes back together—possessing, ultimately, perhaps, its own curious orbits.

Take Mike's thirty years in the world, for instance, and with eleven of those years in Namibia, and that first entire year spent filing papers down in Windhoek, treading water, just itching to get out into the red desert and the world of the deeper senses: the timeless world where even now, eleven years later, he still pauses at nearly every grove of blossoming califraxis he passes and sniffs deeply, and crouches, still, beside the curious and ancient Welwitschia plant, examining with his fingers its coarse heartwood; traces with his hands the broad pugmarks of lions and leopards, when he finds evidence of where they have crossed his desert.

That is a different kind of time entirely, and I am happy that he has found it—he has traveled far to find it—and even being here for the short time that we have, Dennis and I can taste and feel already the qualitative difference in time, or the thing we call time, which may in reality be something else altogether. It's addictive, liberating, beautiful.

The closest I've come to experiencing this same sensation of timelessness, or of a different kind of time, is up above the Arctic Circle. I don't know what the name for this difference is, or the reasons for its presence, but I know it when I encounter it, recognize it by its difference, and it seems to me that Mike does too. And being a creature of habit, I still can't help but try to quantify it, to judge or measure it in the relative terms with which I am familiar. What are eleven years out here really like? Do they "equal" thirty or forty, or even fifty or sixty, spent elsewhere? Or do they pass as if in but a single blink, comparable perhaps to the unscrolling of but a single day?

One rhino, one desert, one *Euphorbia*, one sedge species at the watering hole; one story, the first story, and qualitatively different, possessing that ownership of origin—that originality.

And how nearly the rhino part of that first story blinked out all the way.

It appears, however, that they have passed through a bottleneck, or the eye of a needle in time, and that they received a little push, a little assistance, from the very species, *humans*, that had first led them to the brink of destruction, and invisibility. Now the task is to begin again, to spread forth again and multiply, and prosper. And for this, Mike and SRT have a plan.

Nothing, no rhino, only the day-ago tracks. The sun beats us back to Palmwag, where we wait out the heat of the day, Dennis and I napping in the heated shade beneath the palm fronds while Mike pushes more paper, answers more queries, pursues the meager funding, and assembles and collates more of the trackers' data.

Strangely, it pleases me that we have seen no rhinos, only tracks thus far. It's an odd thing to behold a landscape so lunar, so seemingly without places to hide, and yet to behold none of the giants that inhabit it. Were they small birds, they might be hiding in the thickets, or were they fish, in the greater depths. Any manner of smaller animals can hide in burrows and beneath rocks, but where can the solitary rhinos be? They were here only yesterday, or the day before; they were here only this morning.

It's okay if we don't see one. It's all right—more than all right, it's wonderful—just to be walking across this landscape like no other. It's enough, almost, just to see their tracks and to know that they're still in the world.

It's still hot when we go back out into the desert, but the malevolence or overwhelming-ness is gone from the heat; it's not impossible, merely challenging. We travel to a different section of Damaraland, toward the southwestern corner—closer to those dunes—and prowl the old military roads, the red sand and gravel lanes bladed through the twisted fields of basalt, ghost roads on which no soldiers travel now. We watch for the tracks of those last few rhinos that outlasted the soldiers, and that survived the war. Their world beginning again.

And once again we cut their tracks, fresher tracks this time, a cow and calf, and the trackers spill out of the truck not quite like hounds, but with a greater enthusiasm and intensity than before. It is the very stuff of life, I think, this intensity—the antithesis of sitting quietly, window staring and resting, as a writer such as myself is prone to do, finding myself watching the clock on the wall, waiting for the girls to get home from school and that explosion, that rejuvenation of bright energy hitting the room when they come in the door. They want to go play Frisbee, or softball, or basketball, or soccer. They want to dress and rush off to play practice, or choir, or guitar lessons—the presence of ascending vitality. Here in Africa, we set out after those rhinos with almost that same brightness.

It's true that the trackers are being paid for a discovery, a sighting of the quarry ahead: but there is a hunger, a liveliness, nonetheless, and here too I feel myself reawakened, pulled along, riding in the wake. They move faster on these new tracks, and for a long time these tracks do not wander but instead travel straight toward the next spring; and once again Dennis and I are unable to keep up with Mike and Leslye and Joseph and Himba.

Leslye and Joseph appear to be in their mid-to-late twenties —powerful and young, robust—while Himba could be their grandfather, and is surely a veteran of the times of war, not so long ago. About his own previous relationships with rhinos, and his acquired talent for tracking them, not much is forthcoming, and it seems an indelicate topic to pursue. What matters now is that Himba is working to save them, and that the plan is working, and that he is a very good tracker. He has recently burned his foot in a campfire, so his mobility is hampered; but still, he is hot on the trail, so that I have to wonder what he would be like without his injury. He's a small man, whippet-thin; even tightly cinched, his belt seems incapable of keeping his trousers up. Because the foot injury is too painful to bandage, he is wearing an open sandal on that foot—a cut-out tread of old tire, with leather straps fastened to it—so that his own tracks, interspersed now between and around those of the rhinos, are quite distinctive indeed, as if some force that is half machine and half human has been turned out in search of the rhinos.

We head straight for the spring, though upon reaching it, the trackers now scatter, following the confusion of prints that splay around the perimeter. Here too there are fairly fresh lion tracks, and though the chronology of movement is difficult to establish, it seems that perhaps fearing ambush, or perhaps as a general matter of habit, the rhino did not go into water but merely skirted the dense sedges, looping around the spring in a full circle before deciding to avoid it, and heading on farther south, out across the sere volcanic plain. Though after going in that direction, the tracks loop back yet again—as if the rhino was indecisive—nearing the spring now from the east, though once again veering away, and pointing now—according to the betranced, head-down bee-dance choreography of the trackers—toward one small hill, but then back down the flanks of the steep boulder-studded slope. The rhino paused to chew on one lone *Euphorbia* midslope, gnawed so recently that the poisonous latex is still oozing, and still strongly scented, almost like turpentine, or burning carpet.

The trackers are thoroughly puzzled by the rhino's behavior, and yet knowing that she must be very close, they shift to visual search mode, spreading out into a wider scatter, sailors on this frozen sea of stone casting in all four directions for some sign of landfall—stranded by our unknowingness. *Where is the rhino?*

Here too, on the hill's flanks, large and spectacular quartz crystals abound, the brilliant and beautiful residue of the slow-cooling fire of a heat we cannot imagine—and beyond, on the red plains below, the view is sublime, a blood prairie of stone and late-day golden light.

It is Mike who sees the rhino first. He spies her out in the flat middle of nowhere, with no place to hide, looking even at this distance—a mile?—like a great silver tank moving slowly across the plains, with a tiny silver dot, like a lone electron, at her side. She is shimmering oasis-silver, cruising the red sea like a whale. She is the only thing moving out on that frozen landscape of fire, and to our eyes, she is the only thing at all, animate or inanimate.

We gather around where Mike is crouched—drawn to Mike, and his position on the hill, like electrons ourselves now—and we crouch and stand around him, passing our binoculars and staring at her and her calf, all eyes fixed on them and only them despite the great amplitudes of space all around us, as if in the most divine concentration of worship. And watching her cruise across the desert below—the ridiculous *structure* of her, with that huge body, huge head, huge feet, and huge twin horns—I find myself grinning, smiling ear to ear at the breathtaking elegance of one desert, so immense, and two rhinos amid it, with nothing else around at all.

We sit and watch for a long time, each of us lulled and hypnotized. It is not the horn that is magical or powerful; it is the entire sight, and the idea, and beyond that, the land that accommodates them. We sit as if frozen to stone ourselves, though with our own interior fires glimmering and stirring.

I cannot get over how magical, how *gliding,* seems the procession of the cow with her calf behind her now, again like a ship with sails. There is the profound sense that some larger force surrounds the rhino and aids her in her motion. As if she is being summoned. I have felt this same sense out on the land in watching grizzlies, wolves, and certain other animals, but this is somehow different, here with the rhino and her calf, seen from so far away, and seeming so mythical or surreal, so strangely shaped, and yet so indisputably real and in the present.

Through the binoculars, we can see the individual gusts of dust lifted by the mother rhino with each step, the dust trailing away northward in slanting columns of sparkling mercuric red, columns of dust so well-defined in that late-afternoon light as to seem like spars in rhythmic slant and cast.

It's different, here with the rhinos. There are astonishing messages of sophistry to be found anywhere in nature—in a starfish, in a virus, in a pinecone, in an elk—but this is almost too much. I can understand why the tourists who have been surveyed by SRT are stating that the quality of the experience has nothing to do with their proximity to the animal. Gone from the equation, strangely, is the soft-porn nature predilection, courtesy of our advertising culture and our own hungers within for bigness and nearness. Here, despite the rhino's immensity, the landscape is still a thing we crave. Here, despite there being so *much* landscape —an excess of it, some might even say—we nonetheless crave more, though we crave the rhino, too.

Slowly at first, but then quickly, we stir there on the faraway hillside as if awakened, and a plan develops. We realize that at this distance, we are invisible, despite our being out in the wide open —and it surprises me how quickly we accept and adjust to this new truth, moving around comfortably in the complete anonymity of our invisibility. As if to the rhinos we are already ghosts: or will be, and will remain so, as long as we stay downwind, beyond their ability to know.

We've come through a pass to find her. A little knob of a hill, a tiny mountain, separates us from the watering hole. It's an old eroding volcanic neck, its flanks littered with the scree of its own disintegration, and because it seems to Mike and all of us that the rhino is turning now—again, the movement is like that of a ship just beginning a wide turn, the faintest puff of wind beginning to fill its quartering sails—it occurs to us that maybe she has decided to come in to the watering hole after all, near the day's last light.

If this is the case, we don't want to get between her and the watering hole; but such is the angle of her approach—she *is* turning, is completing her arc now, with the baby nosing along behind her like a tug—that if we can crouch on the flanks of the volcanic neck, slightly elevated, we will be able to watch her pass right in front of us on her way to the water.

It feels a little strange, the six of us striding, hurrying, now almost jogging across the gold and red plains to get to that spot, that unmarked intersection, before she does, with the immensity of her so plainly visible and drawing slowly nearer, and again, with us accepting our own utter invisibility as a simple matter of course. As if we are not seen simply because we do not want to be.

And judging the rate and distance of her and her calf's drift, and of our own, we have to pick up our pace even a little more, to reach that place on the landscape—completely lacking in cover or any other noticeable landmark; simply a blank space on a map somewhere, the near conjunction of two dashed and imaginary lines that will, in a very short while—after she passes—be no more significant than similar dashed lines drawn on the condensate of a windowpane in morning fog, gone completely after first light and not returning.

We reach that spot well before she and the calf penetrate a sphere that might include the distant shapes of us; and still the warm wind is strong and steady in our faces. We settle into our spots like fans arriving almost late to the opera, like theatergoers arriving just before the curtain rises, and with all others already seated, and yet there are no others, the world is empty, and the only difference, the *only* change, between now and seventy million years ago on this one specific dashed-line mark on the map is that six humans have washed briefly across that marking and are resting there for a moment longer.

On she comes, growing not just larger but stronger, and with her tiny calf behind her, hurrying to stay with her—picking its way through the rocks she seems not to even notice.

Our cameras click and beep and whir as we photograph to our hearts' content. The wind remains steady in our faces, swirling our hair. As the rhino draws nearer—still tracking tangentially to our crouched and waiting position, still seemingly headed straight for the watering hole—she grows more and more beautiful. Her considerable muscles are more defined, and the previously metallic silverish glint of her is softening to a cream color. She appears huge, the closer she comes, yet she appears more graceful, too.

The calf—one-tenth her size—looks as wondering and new as is its mother ancient and experienced. The mother's horns are immense, the front one longer than any of our forearms, elbow to outstretched fingertip, and the back one nearly as long; and as they draw nearer still, we can begin to see and understand certain things about their attitude and demeanor, their state of mind (they are mammals, after all, not giant reptiles, and highly expressive), that were not visible to us at a distance.

There is an intelligence to her eyes—why does this surprise me?—and we can see quite easily now, quite clearly, that the mother's overriding concern, perhaps the sole focus of her being and awareness, is the safety of her calf. She is not out for an evening ramble, as appeared to be the case when we first viewed her from a distance of a mile or more, and neither does she appear to be thinking too much about any upcoming draught of clear spring water. She does not look back at the calf, but the cadence of her every step, as well as the swiveling of her large ears, is, we see now, in tune with the calf's movements just behind her, so that we understand she is preparing a way for him in the world.

I have been avoiding the perhaps anthropomorphic assignation of gender to the calf, but I can no longer avoid it: he is acting, in a way I cannot explain to myself, like a little male: goofy, stubborn, muscle-clotted, trip-clumsy, alternately insecure and bold; big-headed, swaggering, oblivious, mother-attached. It's totally unscientific, totally indulgent of me, but that's the word I choose to use for him, *him* instead of *it* or *her*.

I look over at Mike, who is grinning, rapt, and even the aged and hardened Himba is staring transfixed, hunkered on his heels, arms crossed and folded over his elbows, watching the nearing rhinos with a look of curiosity, almost a tenderness, and a keenness that intrigues me, knowing as I do how many hundreds of rhinos he has probably sighted over the course of his long life.

I can guess that he is feeling the same thing we all are—accruing amazement and wonder—as she and her calf come closer and closer. I am fairly confident also that my guess is accurate, for there is an intensity to the movement, a peace and clarity, and a unity, that is all the more powerful for our not-speaking.

Mike, still rapt, whispers ever so quietly, "Okay, no more cameras," and the snick and click of them falls silent, with the last chirp of the single reflex mechanisms whisked away into the wind, with only silence now. And we understand, in that unspoken, unified way, that the mother rhino has just now entered into the leading edge of that sphere, that dome of sight: just beginning to penetrate it as one might begin to penetrate, with one's finger, the outer film of a transparent yet iridescent soap bubble, taut and quivering.

Among us all, only Mike is kneeling—the rest of us are sitting or crouched—and as such, he is raised higher than the rest of us. The wind is blowing his long sunburnt dirty-blond hair behind him, his face aglow partly with the directness of the western sun but also surely with the experience—and his wind-wavering hair is catching that horizontal sun so that it is being illuminated, reddened like alpenglow, making it appear not so much that his hair is on fire, but that there is a force around him, a brightness, a corona.

"Be still," he whispers to us. She and her calf are very near now, only forty or fifty yards out, and after a few more steps, they will be on past us, with their route carrying them on toward that watering hole—and this, we believe, is as close as we will ever be to her. It is very, very close, and it is close enough: more than close enough. I can sense that Himba and Joseph are a little discomforted, no longer quite so lost in the previous rapture of beauty. Himba raises an eyebrow in Mike's direction, and even without Mike seeing this, Mike whispers to Himba, "It's all right" —whispering into the wind to us, not the rhino—and then, just as she is about to pass on by and slightly beneath us, she instead turns and begins coming up the slope toward us.

She can have no idea we are there, yet has swung her own position around and is homing straight in on us: and as she does so, it feels to us as if the myth of our invisibility—previously our power—has somehow been stripped or scrubbed away.

She is not looking at us—she is off by a few degrees—but it's eerie, utterly improbable. Why else is she climbing up this mountain? We go from reveling in her beauty to becoming dimly and then acutely aware that soon we should stop admiring her and shift gears completely, to be worried about our own safety; and yet this is a transition we seem unwilling to make.

It is as when one is awakening from a wonderful dream.

In the way that the six of us — native trackers, Cockney scientist, and American wayfarers — are crouched and crowded silently together, our senses inflamed and astounded by her approaching beauty, and yet also understanding that we are very close to being in deep trouble — that although the issue has not been forced yet, we are at the edge of deep trouble (the white calf has paused some distance behind her, is choosing not to ascend the mountain; why?) — there is something shared and familiar, something ancient and known, about our unity, our fear.

No words are necessary between us, and we remain clustered, frozen, banded together as a whole, yet also each fiercely, independently isolate with our pounding hearts: knowing that if she lowers her head and charges this improbable blank spot on the map, we must scatter like quail, must disperse to the winds. All this is said between us somehow, without a sound or a gesture.

She stops twenty yards away from us. A hundred or more naked miles in all directions, and she has directed herself to, and then stopped at, this point twenty yards from us. And how much more we know in that nearer distance than anything we might otherwise have known or previously observed or considered; things are entirely different, at this distance, in the collapse or vanishing of distance. And though I am certainly not thinking such things in the moment, it will occur to me later that perhaps for much of his life as a child in England, Mike knew this closeness — the unnamable difference of things that might occur, in that final distance. He knew it almost every day, in his visits to and labors at that suburban zoo.

She is not quite facing us directly; she is peering in our direction, chin lifted to test the air, and to present an image of full confidence — but she is off by a degree or two. It doesn't really matter, at this near distance—she is staring at a spot four or five feet to the right of us; and while this tiny wedge of space is certainly meaningless to us with regard to any defense we may soon come to need in the event of a charge, it is nonetheless a scrap of psychological relief: as if we might otherwise wilt or wither beneath the direct or more precise force of her unseeing gaze.

In our new space of fear and beauty, we are able to observe her in a way we would not otherwise have been able to—everything is revealed, here at last—and I find myself astonished by each individual part of her, as well as by the greater whole that stands before us.

Her eyes are filled with intelligence, and with anger but not quite menace; though too, she appears to be possibly considering, or building toward, menace. In our suspension, there is time to note the bristles at the end of her tail, like those of a horse's, and the strange curves and articulations of her bowed legs—in from the shoulder, toward the top of the knee, then sweeping out from the knee toward the ankle, and then back inward yet again at the tops of the huge and all-important, all-durable feet. Out on this landscape, her strange toes do not look strange, but instead, perfect.

The horns, of course, command our obvious and immediate attention—but what I find myself observing most closely about her is her color and musculature, particularly her front shoulders, which support so much weight. I had assumed, from everything I'd always heard, that a rhino was armor-plated, segmented as if with scutes and scales, like a great reptile, even a dinosaur. But staring at her now, I find her to be swathed in muscle, not armor. Her skin appears creamy, with the detail of musculature one might expect in a classic marble sculpture—Michelangelo's *David* comes to mind—and the horizontal sunlight heightens the details of the musculature even while softening the tone. She is brilliant, luminous, and she is made of muscle, not scales or stone.

Mike is watching her most closely of all, in that same mesmerized state of rapture, yet I see no fear in him, only joy and wonder, leavened with concern for the rest of us, in case things turn ugly; and as the mother's musculature tightens and she cants her head that two or three degrees to the left, fixing on us now, Mike senses the coming charge and says, "Okay, stand up but stay together, and back up slowly."

Himba and Joseph and Leslye, having read this same message in the text of the rhino's body, are already rising, and Dennis and I stand quickly, almost falling over in our haste, and the rhino snorts, as her own worst fears are confirmed: humans, aggressors, materializing in front of her as if summoned or produced from some vertical shaft or borehole in the desert, rising before her in an awful blossoming.

She takes a quick and muscular double step in our direction, and beneath the force of the advance, the six of us scatter, falling apart in different directions like so many petals falling from a faded rose beneath a strong gust of wind: and then we freeze, because Mike has frozen—and whether the rhino commanded him to halt, or vice versa (I cannot be sure who was cast to stone first), all movement is suspended again; we are all frozen, there on the volcanic plains, with time flowing around us once again.

We are all hanging there, as if midstride in some child's game of stop-and-go, though it is not a game, of course, but instead a tiny part of the world, a tiny cog and gear, waiting to be decided —and our own hopes for how the matter will turn out seem to have nothing to do with the outcome, only our manners, and the rhino's decision, which might or might not be influenced by those manners.

And so now in that short space between us there is still another thing, mercy—suspended mercy, waiting to be delivered or withheld—and like a dimwit in a chess game, I find myself not leaning overmuch on any expectation of mercy, but planning ahead instead my next two steps: *run,* and then *run again*—choosing my path with my eyes—and yet still, she seems beautiful to me, cream-colored and gigantic in the sun, and with her ribs heaving now in full agitation.

Her calf remains on the stone prairie below her, looking up, understanding that something is wrong but not knowing what it is. The mother tenses, then voids two quick blasts of urine into the wind, the golden spray of each one vaporizing like samples launched from a perfume atomizer on a display counter.

We watch as the mist is wind-whipped toward us, disappearing just before it reaches us, though we can scent, even taste, its bitter acridity—I imagine the metallic toxins of kidney-strained *Euphorbia,* potent within it—and yet still we do not move, and again whether it is because of Mike's spell or the rhino's, I am not sure.

And being atomized like that, while not pleasant, is still not as bad as being gored and trampled, is perhaps not even as bad as being camel-slimed: and seeing that she has voided, I know, without having ever read about such things, that her next step will be either to charge or to retreat.

Her calf is below her; she chooses the latter. And even though our lives were in peril, there is immediate disappointment as she wheels and gallops back down the mountain to rejoin her calf.

Wait, I find myself thinking, *don't go.*

She reaches the calf—the curious, now frightened calf—in a heartbeat. Making a sharp and nimble turn, like a skier sending up a spray of snow, she flares to a stop in front of him, wheeling so that she is between him and us, her huge body back in the charge position, and dutifully, the little rhino tucks in against her flank, not knowing what any of the fuss is about but understanding that somewhere beyond his knowing exists a threat. I can only imagine how frightened the calf must be, for the all-powerful ship of his mother to be behaving in such a fashion. Surely he must be wondering, *What in the world could possibly be more powerful than she?*

We remain where we have stumbled and fallen, the six-leaf scatter of us pasted all over the rocks, low-crouched again, and now to her dim sight it might seem as if we have gone back down into the earth. And like a plant unfolding and blossoming himself, the calf rotates a precise and geometrical ninety degrees to his right, so that his butt is pressed up hard on her flank. It is as military-looking a maneuver as I have ever seen. The mother is facing us, or the place where we were, fiercely, powerfully, resolutely, while the baby—looking east—is staring just as fiercely out at his ninety-degree direction, waiting.

Some period of time passes—we are simply staring, rapt—and then, again as if guided wholly by some older instruction, the calf does a little shuffle-step as some galvanic message passes through him. He rotates another ninety degrees, still flank to flank—hinged to his mother at the hip—and faces due south, while she remains alert, scowling blindly to the north in her own form of knowing-without-knowing. She cannot see us or smell us, but it is her desert, and she knows enough.

Another finite amount of time passes—something precise and measurable, something as perfect and balanced as if by metronome—and now the calf shuffle-steps twice more, almost a skating kind of dance move, fast and happy feet; and suddenly, as if rotated on a dais, he is glaring west, still with the same rigid countenance as his mother, tiny horns upthrust and ready: and once more, they are fastened flank to flank, prepared to defend against all comers.

The same period of time passes once more, and now, with the full 360-degree perimeter having been successfully defended, establishing that no immediate charge by an aggressor is forthcoming, the mother appears to make her decision to vacate the engagement. With no advance warning, she wheels violently and gallops away, and the calf whirls with her, jerked along as if by a string.

Their speed is beautiful to behold. They are pale as ghosts, cream-colored almost to whiteness—but as they gallop into the red haze of the afternoon desert, they begin to glint once more, and seem to grow metallic, to grow mythic.

Nothing could travel across the desert as fast as they are going: not another animal, or a machine. It appears that it is the desert that is unscrolling beneath them, so smooth and powerful is their gait, and I want to know, are their feet merely knocking the stones out of the way, tapping and rolling them aside like croquet balls as they gallop at such high speed—thirty, forty miles an hour—or are they picking somehow, perfectly and blindly, each tiny space where the rocks *aren't*?

Their carriage is so seamless—so flowing—that it seems it must be the latter, and yet if this were the case, would they not even occasionally falter or stumble?

The calf, though game, is falling back; a space is opening between him and his mother, though she does not slow for him, but instead forces him to continue redlining if he is to stay near her: teaching him, *This is how it is done.*

Would it not make as much sense, or even a bit more sense —by a percent or two—to stand and fight, when gifted with such power? In such an equation, the issue of sightedness comes quickly into play. Might not an unsighted animal, or poorly sighted one, be forced to stand its ground in that curious military defensive posture, the half roseate, for days on end, rotating at metronomic intervals to take the measure and scent of the four compass points, but never really knowing for sure, heart racing madly all the while — redlining just the same as if the rhino *were* galloping — and never knowing whether the attacker was still circling or had given up and gone away?

Better then, maybe, to be done with it; to bolt away to some inarguable, ridiculous distance, beyond the cast or ken of lion, leopard, or hyena, and to subsequently spend days, even weeks or months, letting the frazzled heart recover under its own terms, and the land's.

We watch as she and her calf enter that faraway screen of green like arrows fired into its midst. How removed we are already from the real-time flesh and blood of them, the acrid scent and sound of them, the calf's confused mewing, and the scratch of gravel and stone.

It seems to me that we have come through our own narrow wedge of possibility, our own dial of survival rotated but the width of one geartooth, one click to the left or the right — the rhino deciding whether to charge that last twenty yards or to whirl and gallop three miles — and in the freshness of our survival, as well as the beauty we have witnessed, there is among all of us an exultation, a world's-newness, that I find hard now to describe or even fully recall in flavor and tenor, in intensity and tone.

I think that I would almost have to go back out on that same landscape to recapture it — to that precise spot, that one shallow depression on the steep slope of the red hill overlooking the watering hole. I think that I might need to be there at that same sun-going-down-fast time of day, too, in order to have the best chance of accessing that feeling again, returning to it as if in some small loop of days.

It fills each of us, I can tell. We stand there all but speechless, just looking around at the desert and grinning, not even *trying* to articulate what it had been like, but merely continuing, or trying to continue, to inhabit it. The utter improbability of her veering off-course and walking straight up to the invisible and unknowable point of us. Later I will figure out that it was not rocket science but instinct; with our instincts, we had chosen the one best spot for an ambush—even if our ambuscade was to be that of cameras—while she, knowing the landscape so intimately, understood also that that spot was the best place for the ambush of an unsighted or poorly sighted creature on its way to the watering hole. She had spent the entire afternoon checking out all of the weak links and danger spots surrounding the various approaches to her and her calf's own desires and needs—casting into the wind, and then swinging back in a wide circle as if running a trapline: and with her diligence, her perfection rewarded —discovering the threat of us in the last spot examined, as if beneath the last unturned stone.

This, then, is the only marring on the finish of the magnificence of the experience: the guilt of our pleasure displacing her need, and the draining-away feeling of having frightened her to flight. It is a feeling like loss, and it is a strange juxtaposition, there on the tail end of such otherwise joy.

It is easy for me to see how such a thrill could become addicting—how the sighting and tracking of the rhinos could pull a person across the desert in every bit as direct a line as that which summoned the rhinos back to the curtain of distant green. I can catch a whiff, even, of what it might have been like each day, every day, for a small boy in Liverpool: though surely this, out here on the native landscape, is vastly superior, with the senses even more deeply filled.

We wander as if stunned a while longer, and then head back to the truck in violet dusk, holding on to that which has just happened.

"That was good," Mike says. "In all my years, that was one of the better ones." We are all still exalted, but he seems even more so, though quietly. Distanced, somehow, himself.

We ride back to our campsite changed, altered. She was *breathing* on us, she was so close. There was nothing between us and her, and neither was there anything between us and the horizon.

Mike and Himba and Joseph and Leslye confer about the name of the calf. Her mother is called Tina, they tell us, though she also has a Namibian name that translates, roughly, to "She belongs to nobody"—a lonely-enough-sounding name, but one for which Mike seems to relish the mix of romanticism and realism.

Typically, on the rare and joyous occasion when a new calf is discovered, the guide who first finds the calf is given the responsibility of naming it. In this case, it's hard to say exactly who discovered the calf—technically, it could be said to be Mike's duty, as he was the one who, looking up from the tracks, first called our attention to the great shining dirigible, the mother-ship, in the distance, towing her calf along behind her. But by the same token, it was Joseph, Leslye, and Himba who tracked her all over creation, following so perfectly, even when there were no clues, the circuitous dance of her approach and retreat, her labored circumnavigation of the water hole—and it is likely also that all of the guides saw her at about the same time, from their scattered positions across the hills, looking out and down upon the sweep of barren plains below, a slight basin of red rocks, the size of a small shallow sea now drained.

The way the obvious thing, and the thing for which you have been long searching, is finally right there in front of you.

Mike and the other trackers continue to confer. *"Ongoody,"* they decide. The translation means "She belongs to all of us." I was wrong in my assumption that it was a male.

That night, after dinner, still ebullient, we sit around the campfire, talking. We stay up late before finally turning in well after midnight, crawling up into the tents that are perched atop the Land Rover, to reduce the likelihood or possibility of being eaten by lions. The stars are very close, and lie across us like a blanket.

The next morning I wake up in the tent feeling like a child, like the boy I was long ago; waking with the calm assurance—the understanding—that almost everything I would see that day, like all the fine days that had preceded it, would be new, and I would go out into that day hungrily.

Even in first light, the tent is growing heated. The desert pan's first bird is calling, and the air is as still as a bowl of water.

We share a communal pot of oatmeal—a kettle, really, capable of doubling as a cattle trough—and we finish all of it. Dennis laments that our days of steak and eggs and pork chops are gone. Then we get up into the truck and head out, riding it again like a ship bumping over the waves in an ocean of stone. And again the trackers watch the ground, and the horizon, and though it is a different kind of work day, it is a work day.

We strike fresh tracks less than half an hour into the journey, and spill out of the truck feeling doubly blessed: they're from this morning, and we have the full day ahead of us. As long as we stay downwind, we stand a good chance of catching this rhino.

We follow him—his gouges and scrapings in the sand suggest a male—on his own direct path to a different, farther watering hole, taking in again the acrid scent of his piss, the sand still dark and wet from such voidings—and for this animal there is no careful wandering, no exposed vulnerability, only a straight-ahead powerful stride through the world. The luxury of no responsibility; a king among his kingdom, with many subjects. Littering his trail are the fresh deposits of his dung, baled together in loose packaging like that of poorly sorted hay—undigested or semidigested *Euphorbia* branches mixed with a few unidentifiable leaves—and I think again of what a mowing machine a rhino is, an animate bush hog, a force of nature. I believe that such a creation, such a labored adaptation, cannot be an accident, or an incidental response to the process of photosynthesis—no digression, no random assemblage of hydrocarbons—but is instead as much a part of the landscape as the stones and mountains themselves: born of the land and the red dust, as if the shapes of the hills themselves had been one day animated, gifted with life. With the rhino, there is no separation from the landscape, nor can there be.

The rhino is following a narrow little dry wash, the sand and gravel underfoot evidence of at least intermittent flow, probably in the form of a flash flood every few years. (Even a couple of inches of rain, upon this hardened skin of earth, can result in a huge amount of water eventually being concentrated in the penultimate downhill spots in any catch basin.)

A few mopane trees line either side of the creek bed as it

leads us toward the profusion of saw sedges at the spring ahead, and the trees are spaced so evenly, and there is so little other undergrowth, that it feels almost like a leisurely stroll through an English garden.

This feeling lasts only a few moments, however, and is dispelled immediately upon our entering the edges that surround the alkali spring. Mike points to the fresh pugmark of a leopard on the same trail we're following now—the track from this morning, and possibly even fresher than the rhino's, possibly so recent that it was our own approach that disturbed the leopard—and again I am reminded of how utterly without weapons we are. There are rocks all around us, but no knife, no lance, no bow and arrow, and certainly no firearm. For all practical intents and purposes, we are leopard chow.

There's movement to our left, something low and dark moving fast and strong through the sedges—and as it clears the sedges and lifts its head, we see that it's a hyena, and that he's moving away from us, though more with guilt and what seems to be a tightly controlled anger—call it a seething hostility—than any fear. He lopes steadily up the hill, his musculature huge and seeming prehistoric. Again, Africa seems to have chosen to establish the rules all backwards from how they are elsewhere: the hyena's head and shoulders are immense, while his flanks look almost scrawny. The effect makes his getaway lope look even stranger and more singular—as if he is running backwards—and he runs with his head turned back over his shoulder, watching us the whole time, fixing us with a stare that if someone were to describe as evil or malevolent, I might not argue the interpretation.

A second hyena clears the sedges, following the route of the first, and then, farther up, a third, and then a fourth, so that it seems the sedge oasis is *producing* them: and they all run with that same guilty gait, running as if quitting on something, but not abandoning it.

The first hyena, the largest, stops at the top of a small hill and looks back down on us, his odd dark bulk silhouetted against the blue sky: his musculature and coloring not all that different from the blackish-red rocks on which he is standing. The other three animals cross the ridge without pausing, though in all of their gaits, there is that same unapologetic and bristling scorn. It feels like a furious superiority, and the undomesticated resentment of an older, more established resident when confronted with anything weaker. It is ancient and accurate, even if not quite nameable, and then the first hyena disappears over the ridge, joining the others.

Just ahead of us, a small gaggle of pied crows lifts into flight, and Dennis and I are eager to add one and one and one together to make three — to assume or believe that there is a fresh kill just ahead, one that the leopard has made and on which the hyenas and crows are already scavenging — but Mike shrugs and shakes his head and says maybe but also maybe not, that none of the hyenas had blood on its muzzle, and the crows aren't calling quite right for there to be a kill; they seem a little leisurely. He shrugs again. "But maybe," he says, appearing unconcerned that there might be a jealous leopard guarding such a kill only a short distance ahead. *Rhinos*, he might as well be saying, *we're here for rhinos.* And we push on through the saw sedges, following the tracks of that one lone animal.

A covey of sand grouse whirs overhead, flaring their wings and swirling down onto the edges of the shallow pond like a flock of ducks. They water briefly, then swing out onto the rocks, where they become instantly invisible, as if vanishing into the maw of time itself.

A few of the grouse remain at the watering hole, then leap to flight and are gone. Mike gestures toward the departing birds and informs us, almost casually—almost as if to say *Why not?*—that those grouse are often the males, and that they are flying back to the nest with precious drops of water held between their feathers, where they will then crouch over the nestlings and shake those few drops of water into their open and waiting, yawping beaks; that they sometimes fly many miles to do this.

Rhinos, I tell myself, *focus on rhinos.* None of it is any less new or amazing—every plant, every bird, every insect—but this is what I came for, the big things—not quite life-and-death matters, but big things—one of the oldest and most keenly adapted mammals in the world. This is my one chance to observe them. It's wonderful, how my attention keeps getting spirited away by every tiny sight, every tiny movement, as if swirled by a gust of wind—but I need to remember that through this matrix of it all, the rhinos are passing, and we must hurry.

Mike and Himba are as lean as marathoners; Leslye is built like a soccer player, and Joseph a little bit like a linebacker. They are ascending another hill toward a pass, distancing themselves from us again—having taken his draught at the watering hole, the rhino has struck out on what will probably be another three-day journey to who knows where—and either they have lost the trail completely or the rhino is wandering about in an incredibly choreographed manner, or is incredibly confused.

We ourselves loop back and forth, as if tracing the script of some ornate calligraphy, or the outlines of some fantastically complex Rorschach image, and with the rhino somewhere ahead of us, pulling away, doubtless tracking a straight line again, as we grope and wander.

The trail, the invisible tracks—or invisible to me, at least, out there among the stones—leads back to the spring; the rhino decided to come back for a second sip, a second gulp. It is hot, the hottest day we've encountered by far. Mike and the guides hold a conference—we've already traveled a long way—and decide that Leslye should go back to the truck and then bring it up this far, so that the rest of us won't have to double back so far in the rising heat.

We start all over again, following what I trust are the newly hydrated, or rehydrated, rhino's tracks, which once more wander the hillside in seemingly aimless drift, traversing back and forth until they reach the ridge, the pass, opposite of Hyena Ridge, at which point they strike due north for a while before dropping down into the next valley, which contains more grass than anywhere else we've seen: so much more so that it's possible to believe we've skipped time zones, or latitudes, or even ecosystems.

Despite this being the so-called rainy season, the grass is already dry and yellow, wheat-colored, and is growing only in clumps and patches; the landscape is still far and away predominantly basalt rubble, scree, cobble.

How to describe heat? The vaporousness of abstractions, simile, and metaphor fail us—"staggering," or "like being inside an oven" (even these clichés are fiercely true, in Namibia)—and yet insufficient too is the rigid sterility of numerical descriptions. For instance, it is somehow not quite enough to simply state the fact that the ambient temperature on this sojourn is now 114 degrees, with the redrock basalt-bouncing ground temperature easily able to reach 150 degrees. Even a description of the body's responses does not accurately describe the runaway race-heart drumming as the body begins to rebel and protest, if not quite panic; the parchment-feeling of the drying lungs, even within the previously safe and moist environs of the human vessel; the beginnings of a crushing headache, and the disintegration of vision.

For a middle-aged man whose eyes dry out easily, this is a new terrain, a new extreme, and as the day grows hotter, my eyesight grows ever fuzzier, until I worry that it soon may be no better than a rhino's: and it occurs to me that this might be another of the reasons a rhino doesn't rely overmuch on sight in so vast a landscape, and with such optical distortions commonplace. And yet there *are* sight-dependent species here too—adaptations of double eyelids among certain species of antelope, for instance, in which the instruments of survival, ocular acuity and even magnification, are given extra care rather than jettisoned: and yet again, one can witness dramatically the fork in the path, the route chosen or not chosen, with the differences—as well as consequences—hanging stark and dramatic, profound and significant, here.

Part of my vision disappears entirely—the lower curve of my left eye rimming with darkness, and my right eye dimming, and my head throbbing—and I sip more water and stumble on, watching as the trackers glide between and among the rocks like spirits. There is no visible shade in any direction. Mike stops and points out yet another blossom, another loner—once more, such a different strategy in this other-world. Long gone, or never-were, are the large pastoral fields of England and the rest of the newer world, in which pollinators are courted, coaxed, summoned by vast displays and profusions of flowers.

It's a large cactus-looking plant, lacking spines, but possessing that same green dull-waxed look, the varnish or wax protecting the plant's interior moistness against the great heat beyond. The blossom is large and white and kind of raggedy-looking, like an old lace handkerchief tossed out the window of a train and left to age in the elements. It's so thin as to be translucent, and with the first frays and tatters beginning to show already, here near the end already of the official rainy season. A quick burst, a one- or two-day marshaling of moisture, and then the subsequent expulsion of beauty, the manufacture of a momentary beauty, and a divinity or power of odor that no insect, whether bee or beetle, fly or ant, can resist—but then, already, the fading.

"Smell it," Mike reminds us, with the pride of a sommelier, as if we have traveled all over this vast estate in search of this one specimen, this one fading vintage of scent.

My near vision still works; it's just my faraway vision that's going away in the crushing heat. I have no idea why, and have not mentioned it to anyone, for what is there to do but hope that it gets better with rest, once I am back in the shade?

The scent of the blossom is not entirely sweet; it is not per-fumey, like a rose or honeysuckle, though neither does it possess the fly-summoning odor of carrion, like some of the desert species Mike has pointed out, and which, via that strategy, contain their own kinds of success.

Instead, it is earthy, almost smoky, like the tanned hide of elk or deer, or like the scent of a kitchen, or an entire home, after a fancy meal has been cooked—garlic and browned butter and sautéed onions and shallots, soy and paprika and cumin, and meat roasting in the oven. Vegetables in simmering, salted broth, and bread baking.

We linger at the lone blossom, taking our time in its com-plexity—the blossom flutters in the breeze, frays against itself some more; it will not be here much longer, but no matter, its work is done—and then we rise and troop down through the pass and over to the other side of the ridge into the next valley below, the grassy valley.

And in the grassy valley—smaller, nestled between a series of hills in such a manner that perhaps the scouring wind is unable to export the slow products of erosion, allowing a faint layer of sediment to be retained—enough for a dab of grass here and there—we finally spy the rhino, again a mile or more distant and moving directly away from us, heading straight back into the wind, and moving at a good pace—walking steadily, purposefully.

If possible, the sun is beating down on us hotter than ever—again, the impression is that by crossing over into this little valley we have stepped into an alternative reality, or another band or horizon of it—as if such things might be stacked horizontally, like sedimentary strata—or like another planet: rhinos on Mars, humans on Jupiter.

We stop and view the back end of the rhino moving away, but not for long; even at this distance, we can see how single-mindedly he's moving, and it doesn't take an advanced degree in rhino behavior to know that the sun is baking him as it is us, that he's hurrying to get to some shade, any shade, in which to nap—and we are chagrined to see, in the farther distance, a single *Euphorbia*. It probably doesn't cast quite enough shade for a rhino as large as this one, and in such heat, but beyond it, there are two perfectly shaped small mopane trees, each tree's canopy rounded like a lollipop. The two mopane trees are about forty yards apart, and if the rhino chooses either the *Euphorbia* or the first mopane tree, we'll never get to him in time; but if he chooses the second mopane, we might have a chance, if we hurry.

The trackers before me, for instance, are transformed suddenly, magically, into marathoners: as if, having glanced down at the ground, I look up to find that in that half-second's or second's skip of time, we are in a marathon, and that Mike, striding powerfully across the broken plain, long brown-gold hair flying, is in the lead, though with Himba and Leslye not far behind, flanking him as if in a wing, and with Dennis and me suddenly much farther back; and I lean forward into a run, as if trying to cross that fractured plane of reality, and try to catch up.

To see them hurtling across this landscape, and hurtling also straight toward, rather than away from, so formidable a creature —running toward a dinosaur as if to embrace it—is an odd reordering, a wonderful sort of backward logic breaking through.

Mike and the trackers are closing the distance, with a slight angle on the rhino, who walks right on past the *Euphorbia* without even browsing, intent now on sleep, only sleep. I remember Mike telling us earlier of how, in cooler weather, the rhinos will lean over against the *Euphorbia*, crushing it down with their weight and settling into it as if it's a hammock, so that they float a few feet above the ground, suspended in the myriad supple branches, and with slight breezes stirring beneath and all around them, bathing and cooling them in the manner of the baffles of a radiator; and of how sometimes the rhinos nap so hard in this repose that Mike has believed some of them to be dead and has walked right up on them, too close, to investigate.

If Mike and Himba and Joseph are marathoners, displaced somehow, in the blink of an eye, from the urban canyons of some big-city long-distance run—the Boston Marathon, or New York's—then I think that in this alternate reality I am but an actor: Humphrey Bogart, perhaps, in *The Treasure of the Sierra Madre*, for my pockets are so stuffed with crystals I've been finding that it's difficult to run, and I keep lagging farther behind, with the rocks in all my various shorts pockets clattering and trying to sway me first one direction and then another, as if I'm running with a dozen different bags of water strapped to me, all aslosh;

and as I run, I'm having to grip my waistband with one hand to keep from running right out of the freighted shorts.

Mike is gaining on the rhino now—we've run a full mile, I think, at a six-minute clip, which is pretty good, in these rocks and heat—though because we're chasing a moving target, the rhino is still a long way away from Dennis and me, receding ever and always, like some African version of Fitzgerald's tide—and with more chagrin we can see that the rhino has paused, is lingering at the first shade tree, where he seems to be studying it; and if he lies down there, we won't be able to move in close to him, for fear of disturbing his much-needed rest.

Mike and Himba and Joseph are close enough to him now that they have to slow to a hunched-over walk, to keep from highlighting themselves against the skyline. Is the rhino resting, or preparing to lie down? His pause is allowing Dennis and me to catch up, and we finally gather near Mike and Joseph and Himba, where we can watch the rhino, which is suddenly close after having been, all day, so far away: and once more I'm struck by the nakedness of things; by the way we're just kind of hanging out there in the middle of space, with no defense, no guarantees of safety.

Like the needle on some free-spinning dial, the rhino could easily rotate 180 degrees to where he is pointing right at us, and could then proceed directly toward us: and there is nowhere to run, nowhere to hide. It is a little like walking right up to the grinning skull of death, I think, and observing it unobserved, except that it is so profoundly beautiful.

My heart is beating, though not so much from the run or the heat, but from the other thing. We stand there and watch as the rhino tries to make up his mind: weighing his options like a judge. We sip from our water bottles. The guides never travel with water—I have never seen them drink water before noon, though we are fast approaching the noon hour. Being an eternal worry-wart, I wonder if this time maybe we haven't overextended ourselves a little.

And yet it seems not to really matter, for here we are again, connected once more to our heart's desire: within full staring range, full breathing range, of one of the surviving rhinos: an old battle-scarred veteran whom Mike, Joseph, and Himba recognize as a male who can sometimes, Mike informs us, be a little testy. And as such, they are playing off of him, the three of them indecisive about whether to go closer.

Does the rhino sense us standing behind him? It seems now that he might be considering something other than mere shade; that he is considering his larger world, and intuiting, somehow, this new intrusion into it. The wind is still very strong in our faces, a heated blast of gusting breath, and yet it seems to me that now there is suspicion in his pose.

He stands staring at the tree's inviting lozenge of shade for a long time—what passes through the ridges and ravines, the overheated bafflings, of his mind, in such a reverie?—and we in turn stand for a long time staring at him, the furnace breath of the desert whipping at us, not as close as we were yesterday to Tina and her calf, Ongoody, but close; and given the history of this particular rhino, as close as we need to be. Close enough for Mike and Joseph and Himba to be concerned and cautious.

Finally, the rhino looks away from the shade and proceeds farther south; and we surge forward, following him in single file, using the unselected tree as a screen, should he stop and gaze back into the nothingness of his suspicions, if not quite his fears: for it is all but impossible to imagine this animal being afraid of any other creature, or even any other force, on earth.

Perhaps ten or fifty or even a hundred million years ago, there was something to fear. But today?

Perhaps the last hundred years—the onrush of soldiers with machine guns—has occurred so fast, relative to his evolutionary scale, as to seem like but a single brief dream, a nightmare from which he will emerge at any second, blinking.

He walks slowly now, traveling a trail worn smooth to the second tree, which he might or might not be able to dimly see. Maybe he was remembering the first tree as being not quite as comfortable as the second, or calculating, in his ancient way, the current time and temperature, and the sun's drift, in such a way as to consider the future, and to plan ahead, placing himself in the best spot to avoid the coming heat of midafternoon.

The rhino is almost to the second tree now. Mike whispers to Dennis and me to stay put by the first tree—there are too many of us to approach any closer—and like bandits, Mike and Himba and Joseph half run and half tiptoe out into the broad light of complete openness, trotting along right behind the rhino—woe to them if he should happen to turn around, for they are now well within his sphere, his bubble of sight, but the rhino does not turn around, and instead plods resolutely toward that second tree.

The trackers need to get just a little closer to get an absolute identification, based on horn and ear descriptions. They're all but certain this is the charger, but they need to be sure; and they need to close that final distance before he lies down for the rest of the day, and they need to get a transverse view of him as well—which means, of course, that at this distance, he might be able to get a transverse view of them.

A pied crow appears from out of nowhere, basalt-spawned, and alights in the tip of the second tree, beside which the rhino is now standing; the rhino is almost ready to drop to his knees, almost ready to call it a morning, checking out until dusk, or even the cool relief of starlight.

The crow—mischief-maker, trickster, and, who knows, perhaps desiring a little havoc, which might eventually play to his scavenger's advantage, whether ending poorly for the men or the rhino—begins to scold furiously, a sound so shrill and agitating amid the otherwise-seeming lifelessness of the windy plain that it is like a kind of fracturing, as abrasive and startling and disconcerting as the breaking of panes of glass.

The rhino tenses, and the trackers—clearly and cleanly busted —also tense, then drop to the ground in the lowest of crouches, hoping that the waving fronds of grass will obscure them when the rhino wheels in a half-circle to examine his backtracks. The rhino squints in their direction while they remain frozen, and above them all, gleefully, the crow continues his merry and energetic song of disruption, calling out to man and beast alike, *Look, your world is falling apart, can you not see it? Look, look harder—it is right before you!*

Nothing is what it seems, all parties know it, danger is everywhere, and yet there is no evidence, or rather, the evidence cannot yet be seen or is not yet in motion.

The crow can see it, and the crow is shouting it.

This time the rhino is not staring in considered reverie, but is instead drilling a borehole into the present, staring out just over the tops of the trackers' heads. It seems certain to me that if he sees them—if he spies even the slightest little movement—he will charge, will uncoil his fury and what is now surely his fear, for now he is a fierce and perfect study in supreme pissed-offedness, with the dream, the nightmare of man, continuing.

There is no doubt in my mind that if this rhino could see

his target, he would charge with relish: and not a mock charge such as we saw yesterday, but the real thing; that he would see it through.

He huffs, then trots forward briskly anyway, as if anticipating his suspicions, his pursuers, will materialize before him — almost as if commanding them to do so — and when they do not, he flares to the south and gallops out into the heat, seemingly as much to distance himself from the ceaseless and maddened scoldings of the crow as to escape any former troubles; and once out in the heat, he stops again, tips a hoof back as if striking a bodybuilder's pose, and there he awaits whatever trouble, whatever challenge, might be coming his way: supremely confident, and still supremely pissed.

Down in the thin and not very tall grass, Mike and Himba and Joseph remain huddled like birds. The rhino continues to glare, but his scowl is somewhat diluted because he doesn't know which direction to address. To maintain his regal bearing, it seems, he must choose a direction — that crow was fussing at *something* — and so he continues to stare furiously northward, insistent on his magnificence before an audience that even he is beginning to suspect may not be there. Hell, maybe the crow was fussing at *him*.

Fifty-five million years of finely crafted selection — or more simply, too much pride, to return to the crow's tree. He decides it is better to play it safe, and the correct choice is always the harder one.

Betraying no weakness or ambivalence whatsoever, the rhino quits the scene and begins walking northwesterly — back toward the pass, and toward the spring — at his regular pace, as if it had been his intent all along to leap away from the shade of that tree and venture back out into the superheated red pan of the desert.

And once he has quartered away from them and has exited the forty-yard bubble of vision, Mike and Joseph and Himba arise from their paltry hiding spot—delighted, I am sure, to be alive—and reconnoiter with us over by the first tree.

"That was very nice," Mike says. It wasn't quite the quality of the sighting yesterday. For one thing, we had to run hard to reach this rhino rather than sitting down and experiencing the one-in-a-million chance of having the rhinos walk directly up to us, but still, it was a good sighting. There was satisfying proximity—the thing Mike lives for, I think, in a fantastic and beloved landscape —and there was even a little buzz of adrenaline, which I think may be another of the things that pleases him and makes him be more alive.

The mountain biking, the surfing, the rhino-charge boulder-hopping, and the old schoolboy days of rugby, back in England: there is nothing about his gentle spirit and demeanor that suggests he would particularly pursue these edge activities. Rather, there is a sixties-era kindness and quickness, a kind of diplomacy in the world—again, I recall his mantra, about which the students tease him almost daily: "A Good Heart Never Fails."

Yet it will occur to me in retrospect that neither is there anything about his nature that will cause him, like some shrinking violet, to shy away from any of life's riskier and more robust activities. And with the precision of hindsight, certain moments not particularly noted or signified by me will, in subsequent analysis, reassemble themselves into the true or other picture—the proverbial *It should have been obvious all along* reaction.

Take, for instance, the post-viewing satiety—the high—that is emanating from Mike right now: eyes aglow, face exuberant, a spring in his gait. The relief and release and joy and awe all bundled together, the moment of life made manifest.

"If we hurry," he says, "we can get out ahead of him again. We can get some more pictures." A glance at the sun. "It's on the way out," he says.

My vision is better, having stabilized in the shade; it was just some buildup of protein and plaque on my contact lens, due to my eye ducts drying out in the heat and wind. I'm good to go again, and so is Dennis.

We turn and hurry once more to catch up to the rhino—to draw closer once again, as if already unable to accept any distance —addicted to, and craving, being within that forty-yard sphere in which the new world vanishes, and a far more ancient world re-emerges, and is known, or remembered, or discovered.

Because the wind is at our back now, we have to parallel the rhino rather than follow him in a direct line; and because of his steady and powerful pace, we find that once more we are having to run to catch up with him. Drawing even with him, yet flanking him just at the edge of that forty-yard cast of rhino-vision, we pace alongside him, the rhino as immense as we are tiny. We adjust our pace to his, calibrating our steps perfectly, until it is as if we are but unseen shadows; and it is as surreal a sight to me as I have yet seen, the five of us strolling alongside the rhino out here beyond the middle of nowhere, as strange a sight to me as if we were to be walking him down Park Avenue on a leash: attached to him by an imaginary tether, and with the landscape and the creature himself so gothic and fantastic that it seems it should not be the coterie of us attending him on this walk, perhaps. Not Himba with his mismatched flapping sandals barely covering the bottoms of his feet, and Dennis with his propeller-chopped arm riding slightly crooked in the heat, and me with my ridiculously sagging shorts, my rock-laden pockets, but instead, some glamorous fashion model, with a leash glittering with sequins, flashing out here beneath this end-of-world heat, and diamonds, as the song says, on the soles of her shoes.

Heat-struck hallucination, or simply the imagination stimulated by an incomparable landscape, and an incomparable species? I don't know, only that it is not like anything I have ever done, hurrying to stride alongside this big cruising creature, an animal that could so easily dispatch any or all of us, should we breach that vague and unknowable and perhaps wavering line of sight. Nor have I ever traveled through such a red amazement of landscape, such all-other barrenness. What hallucination, really, could be that much stranger than the heated reality of the moment?

We understand that the rhino is heading back for the pass through which we all came—the pass leading back to the leopard's and hyenas' watering hole—and so after snapping a few more photos, we surge ahead of him, being careful not to cast our scent onto his backtrack. We are barely able to get far enough ahead of him—the forty-yard distance—to position ourselves just downwind of the pass, the slot through which we are certain he will pass: as if it is all but a game of checkers being played amid the two simple variables of wind direction and distance of sight—those two variables as simple, in some ways, as the red and black squares of the checkerboard.

Once again then we find ourselves crouched and waiting, photographing wildly, as a gigantic beast approaches us unknowingly. We crouch, hidden only by our motionlessness, and watch like voyeurs as the giant strides on, coming almost directly toward us, powerful and unknowing.

And again, the stakes are raised substantially once the rhino enters the forty-yard sphere we have established with our temporary encampment. In the here and now, there can be no turning back, not until after the rhino has passed, and so we sit and wait and watch as he draws ever nearer and grows ever larger, the tiny grouping of us once again vulnerable out on that desert floor and before such a behemoth.

Himba—the oldest, and the one with the most rhino experience—is the first to begin to fidget, muttering something to Mike, and to all of us, under his breath, and backing up ever so slightly and positioning himself to leap up and run. He places a hand lightly on Dennis's and my arms, suggesting, counseling, that we back up with him, but Mike shushes us all, whispers to stay, but too late: the rhino stops just before entering the pass, which is now slightly to his left, and turns his head to squint in our direction, slightly to the right.

He scowls and peers for a long time, and then takes two curious and completely unaffrighted steps forward and tenses.

I cannot tell if it is anger that begins to fill him or not. I have no idea, can find no clue.

The rhino continues staring at us. I think that he is close enough to see us—maybe even close enough to see us clearly—and I long to be back slightly outside that forty-yard perimeter rather than just inside it. I can tell that Himba, beside me, feels the same way.

The heat is terrific. The wind is in our face again, but the rhino is looking right at us.

He lifts his tail like a plume and sprays a squirt of urine into the wind, followed by a second blast, and once more we can smell pungent barnyard scent. He watches us, or the direction of us—not a camera dares to blink, despite his dramatic pose—and perhaps because of his greater size and dominance, or because he has nothing to defend but himself, he appears less threatened, less frightened, than Tina was yesterday—tough with that extra confidence, even more menacing, as if now relishing the possibility of engagement with puny desert-stranded humans.

I imagine that beside me, Joseph and Himba surrender a little bit of hope — as if the jettisoning of such baggage is, in their long experience, one of the first of a series of coming necessities requisite for survival, with subsequent steps surely involving nothing more complex than jumping up and running. As if so finely tuned might be the matters of survival here that even the carrying of so vaporous and intangible a weight as some tiny hiss or whisper of hope might be an extravagance.

Is this how we die, then — not in the presence of the familiar, or upon the foundation of all that has been learned and known up to that point, but instead upon some stratum, some plane, entirely new? We seem to be at the edge of such a place now — in the irritable, unbearable heat, the rhino seems to be the size of an elephant — and we sit there, pinned by the sun and by the rhino, and by the confluence of our enthusiasms taken to their further boundaries, for is this not what we sought: to see a rhino?

Again, miraculously, the rhino chooses not to charge — chooses disengagement — and turns quickly to the left, trots up and over the pass, then settles back into a walk — partly to accommodate the heat, I think, but also not wanting to give us the dignity of any further awareness. He is a king, and he proceeds on toward the watering hole as if that was his desired and primary destination all along.

We rise from the stone sculpture of our stopped-in-time hiding spots and visit, enthusing over our survival. We will not follow him again — or rather, we will hike out the same way we came in, but will not pursue him any farther, will not risk any additional disruption to his equanimity, his thermal maintenance, his *life* — already we have transgressed — and we'll find Leslye and the truck and head back to camp, to hide out from the sun's broil for the rest of the day.

We skirt the spring by a wide berth, side-hilling far above it to keep from risking yet another spooking and subsequent displacement of the rhino—we spy him far below, headed for the shade, the burrow, of the sawgrass, and I hope that he will wallow in the spring, where the evaporative action of the wind will then carry away his body heat, as if pouring water onto and then turning a fan onto a heated skillet—and I have to say, were there not the lion and leopard factor going on down there, the idea is pretty inviting for ourselves as well.

We've all overextended ourselves and are low on water. Dennis's is gone entirely, Himba and Joseph never carry any, Mike has about half a canteen left, and I have saved two inches for when we get in sight of the truck, or in case something happens between here and there.

We continue on past the spring, following the wandering script of oryx and springbok trails, which, though linear, possess a slightly spidery scrawl to them, like the seismograph-wiggle in the handwriting of a very old person: trails worn smooth to the bedrock among the all-else scree as if swept with a broom, and far more ancient than any road or path of man—the slight squiggle in them arising, I suppose, from where the stones and pebbles that once rested there were hoof-nudged, inadvertently, to either the left or the right—and as we cross each ripple and ridge of the side hill, traveling farther back into the old valley up which we walked earlier in the morning, we expect to see Leslye and the truck, parked and waiting for us, as was the plan.

No such sight awaits us, however—indeed, we had understood that he would drive all the way up to the spring—and since it seems that we'll need to hike all the way back to the truck, the question seems only to be whether there was miscommunication, or whether there are mechanical problems with the truck. In our own heat-stricken state, it takes no imagination to consider a boiled-over radiator, with the iron machine having quickly grown at least as fevered as we are, and unable to go any farther.

We'll know in a few miles. It's not as if we've been driven out into the desert and been dumped. There's scads of water back at the truck—sure, if the radiator's sprung a leak, that might complicate things somewhat, but there's enough at least to get beyond the moment. Troubling thoughts stir my consciousness, such as how one of us might travel the thirty miles back to base camp to get a replacement vehicle, but these thoughts do not hold or attach. That is in the future, and the heat is too overwhelmingly present in the now.

It is a strangely liquid heat now, flowing and rising up around our ankles, particularly, as if the basalt has reverted—has been melted back—into molten lava. Curiously, and most unpleasantly, I can feel a semidistinct gradation in the heat that is rising up from the basalt with that extra and more intense heat about calf-deep, so that it feels remarkably like wading through a heated liquid, even as a different kind of heat, and one more debilitating, shines down upon our baking brainpans, and our skin parches and wrinkles like that of old lizards.

Ahead of us—distances beginning to separate once more, as if a string that might once have connected the five of us is now beginning to fray and tatter—Mike shares the last of his water with Joseph and Himba. They are so hot, and have been so long without any, that they actually accept it. I offer my bottle to Dennis, who looks at it and shakes his head, gestures to me, and I take a sip, cutting the two inches to one.

We stride on, feet clunky and heavy now—the rocks in my pockets continue to clunk and rattle and sway catawampus, and still I must grip my waistband to keep from walking out of the crystal-laden shorts—and over time, the five of us spread apart laterally as well, so that seen from far above, our movement would resemble that of motes or debris drifting in a larger current, traveling not by their own power but embedded instead within the directive of a greater force: like rafts or leaves that have floated far downriver and emptied out into the splay of a delta, and then out into the spreading ocean that has received them.

I hear a loud exhalation behind me, a curse, and turn quickly just in time to see Dennis's ankle fold, pitching him forward and down the side slope, the dry gully, he'd been descending. He bounces heavily when he hits—his pack is filled with camera gear—and he rolls like a barrel about fifteen feet down the slope, rolling across some rocks while scattering others before him—rolling like a log—and when he finally reaches his angle of repose and sits up slowly, his face is pained, and bright blood shines on his knees and shins.

It is his folded ankle that hurts worst, and he sits there for a moment gripping it and grimacing, rocking in pain, but knows better than to unlace his boot, which would allow the ankle to begin swelling so much that he wouldn't be able to get it back on.

The best thing of all for it would be to soak it in cold water—to sit at the spring, perhaps, for a day and a night, barefooted and bare-ankled, cooling the fevered flesh and waiting and watching and listening, watching the sand grouse come whirring in to splash and sip in the shallows, before wheeling back out into the dusk, ferrying the day's second load of water back to the babies. To listen to the lion's cough and roar, and to the strangling wail of hyenas moving down off the ridge, coming closer. To hear then the splashings and gruntings of very large animals—rhinos, elephants, even giraffes—all summoned somehow from this lifeless stone.

Unfortunately, there is nothing to do but walk on. Mike and Himba and Joseph have already disappeared around the next bend—we imagine that they have paused only a shorter distance farther on, where they are gathered, staring down at the parked truck below, with Leslye napping, perhaps, in the front seat, both doors open for ventilation, his feet sticking out of one end. I take Dennis's pack full of camera gear from him and help him up, and he hobbles on, blood-streaked and gimp-ankled among the stones, trying not to fall again.

But around the next bend there is nothing, only absence— neither truck nor travelers—and with a dim though deepening concern about the many different ways things could begin to go wrong, we travel on, ever heat-wearied, ever wobbly, ever frail, and with the knowledge of that frailty informing us more fully than it had even a few minutes ago. A desert education.

At the next bend, too, there is nothing. What the hell? Have they broken into a maddened sprint? Has a lion carried away all three of them? At the next bend, finally, we find them crowded together beneath a lone cluster of thorn bushes, staring at the spot where the truck used to be.

Dennis and I all but crash into the shade, skating just along the precipice of heat stroke. Mike and Himba and Joseph grunt and rearrange themselves slightly to allow us to accommodate, as if in a game of Twister, the volume and mass of our boiling bodies to the irregularity of faint shade cast by the bushes. We bend and stretch, shaping ourselves into that crooked shade as if reassembling the jigsaw pieces of some fractured puzzle, and we lie there for a while, panting like dogs, brains boiling, trying to think things over.

Mike tells us that Himba and Joseph have already gone out and examined the truck's tracks, have followed it up the road and back—the trackers tracking the tracker—and found where Leslye drove up the road a distance, then for some unknown reason turned around and drove off in the other direction.

All we can do, it seems, is wait for him to come back—to come back with water—and to hope that he is not brush-stuck or gas starved or caught on some jagged rock, or radiator busted, or differential jammed, or bearing locked, or anything else like that. We sit in the miserable heated soup of what passes for shade —of what, on visual inspection, might pass for shade, but is lacking in what I recall are the traditional benefits and characteristics of shade—and in our anguished leisure, we scroll torturously down the long list in our minds of all the other various ways a vehicle can stop running, though we give voice to none of these suppositions, petrified each, I think, by the power of a jinx.

Our water is gone, all of it. In this manner, we are exposed to hang by a thread, though strangely, despite our vulnerability having been made imminently more manifest to us, I grow slowly more comfortable with these newer, even if ferociously bounded, parameters: an alteration if not collapse of time.

We don't know when or even if Leslye will return, but with our new constraints so fiercely adjusted, I find it interesting how quickly I come to understand that patience, and the adoption of a kind of non-waitingness, is the best way to wait. He'll get here when he gets here—or not—but the sun, as well as our frazzled condition, absolutely precludes any farther explorations.

Instinctively, and with surprising calm, I discover that the tendency in such a situation is to make an inventory: water, three miles distant, and shelter, shade, right here, right now. Some matches would be nice, for later in the night, but we don't have any. Perhaps we can make a bow drill out of a bootlace, if need be. Food is not even on our radar right now—too stinking hot —and so we lean back and rest, and watch the horizon without really watching, and consider the only real variable we have left to manipulate, *shelter.*

We rest a while longer, then rise and scatter, each of us seeking a slightly better shade bush, one that each of us can have all to ourselves, so that we may each crawl all the way up into the cooler core of shade, nestling in against the trunk and letting the slow sundial of the day work its way around us as we remain motionless, rather than constantly rearranging ourselves, chasing the irregular perimeter of that sundial while at the same time repositioning ourselves so as to also accommodate the skewed positions of our shade-mates.

In this deeper shade, separated now but still each within sight and calling distance, we nap.

The water is only three miles back. What a luxury; what an indulgence. What would it be like to really be stranded without knowing where the water was—the needle in the haystack— and to not have a truck already out searching for us? Again, I am surprised by how reassuring, rather than alarming, I find this affirmation of our frailty, and our reliance on grace, mercy, luck. It's nice in a way, this not having to any longer pretend we are sturdy or even durable in the world.

And yet, such is the century, and the tiny space reserved for such rarities, that less than an hour passes before we first imagine that we hear, and then really do hear, the distant growl of the sole internal combustion engine anywhere in existence within, say, the fifty-mile radius of our hidden, panting position: the truck growling its way toward us, returning to us like an iron filing to a magnet, or vice versa. How quickly we rise, donning once more and easily, immediately, the mantle and cloak of our old-new ways, our addictions and weaknesses: our truer and perhaps more lasting fever.

Leslye had been playing hopscotch, was all: reading our tracks and the rhino's, and trying to position himself between and within the various little valleys in such a way as to not spook the rhino by parking downwind; and he had been trying to guess where we, and the rhino, would go next. He had even found our tracks from where Himba and Joseph had been tracking *him*, and had been able to piece together what was going on, though he had been lagging just a bit back in time: the tracker following the trackers following the tracker.

The rest of the afternoon is spent back at camp recuperating, shade-lolling in our breezy tents pitched atop the Land Cruiser —the dense scent of sun-heated canvas like that of some strange and narcoleptic blossom—and with no more meaningful activity than resting and reading, the day seems to hang, as if stuck on some sort of high center.

Just before the beginnings of dusk, we go for what is more of a recreational drive than a scientific fact- or track-finding expedition. We ride in the back, searching, but there is a keenness and intensity missing. We find ourselves scanning the horizon, mesmerized by something—the residue of peace?—staring out at the same spot on the horizon long past the point where we have ascertained there are no rhinos visible in that vector, staring-without-seeing, in a leisurely way, which is every bit as delicious and relaxing a pastime as the day's earlier waiting without waiting.

We find no new tracks, though we know also, as the ground unscrolls beneath us, that we can encounter them at any moment. Even when we are not really looking, they might yet appear, looming, as if revealing themselves.

As we ride, Mike's asking me about my family, and where I live: about my life. He's intrigued not just by the domesticity of my life, even amid a wild landscape, but by my embrace of and desire for that domesticity. And he's fascinated too by whatever arrangement I have made with my wife—as curious that she should find the wildness that surrounds our life of domesticity as pleasing to her as is the domesticity embedded within that wild landscape is attractive to me. One of the rarest of things—the just-right fit—achieved not across millions of years but within the hurry-up, one-shot-is-all-you've-got span of a single life.

He asks about what the children think about such things, and how they get along, living so far from the rush of the world, and marvels again that I've been able to find a partner for whom such an arrangement works. He looks out at the reddening horizon, clearly considering the possibility, or perhaps rather the improbability—perhaps not quite an impossibility—of enticing, or discovering, a partner willing or even desirous of inhabiting this all-but-lifeless stone field—*Perhaps in another life*, he might well be thinking—and we ride quite a ways in silence then and watch as the sky deepens to violet, and the cold stars come out and around and above us as if released from a place of previous captivity.

"That party we had at the end of the semester last year was fun," he says dreamily, a little further into the night: the red road glowing in the truck's headlights, the truck bouncing and jouncing, heading back to camp. He'll miss the celebration this year

—he's going to accompany Dennis and me to Rhino Camp, the luxury tent camp that lies a couple of hours from here, which is run by his friends Chris and Emce.

"Everyone showed up," he says, remembering last year's party —friends and strangers alike, all showing up to dance and play music out in the desert deep into the night. "It sure was fun," he says, dream-sounding again, and reflective, and I misinterpret what he is saying, and I suggest to him, "You should go this year. You've already given us so much time, more than we could ever have hoped for."

He shakes his head. "No, that's not what I meant," he says. "I'm just saying it was fun." As if marveling, to himself, at how *much* fun. Like the desert's one inch of rain per year, perhaps— the anomaly, rather than the thing that defines or identifies the desert. He smiles. "There'll be other parties," he says.

Part III

DUST

BACK AT PALMWAG THE NEXT DAY, WE SHOWER IN THE
cobra stall, rinsing quickly as much to conserve water and solar
battery power as to avoid any snake visitation, and then we lunch
at the lodge next door, a beer and a burger, shade, and the coo-
ing of doves, with their sleepy admonitions or mantrap of "Work
harder, work *harder*." Was it only this time yesterday that we were
all but crawling on our hands and knees across the stones, lost
and abandoned? Is this how the world really is, beneath the façade of our clocks, when we throw away the tyranny of clocks
—filled with anvil heat and anguish in one moment, and relief
and pleasure in the next—traveling through and among and between such emotions and experiences, not like anyone with a
plan (much less any semblance of control), but more like a leaf,
wind-whirled?

While we were gone, Mike's ancient Land Cruiser has been
returned to him, having required one major repair or another—
he's been separated from it for weeks—and he's happy to finally
have it back. To me it just looks like an old truck with a high
platform welded atop it, but I come to understand that Mike
views the vehicle in a slightly idiosyncratic but understandable
way, as one of his partners in freedom and possibility: durable and

dependable, and if it doesn't quite have an imagination, neither does it have the ability to discourage or criticize or in any other way impede Mike's explorations and investigations.

Indeed, later in the afternoon, as we are loading the truck for the next leg of our journey, he will walk up to it as if it is an animate object, slap his hand on its fender, and exclaim, once we've got it fully packed, "Now *that's* a truck!"

Before the truck is loaded, and before Mike breaks away from paperwork, and away from the campaigns and schemes and hopes and strategies that are pulling him in so many directions—rhino tourism, community empowerment, foreign press publicity, scientific data, park establishment, councils and committees, training of volunteers, protection of existing habitat, with maybe even a partner out there, a partner who craves and loves the solitude and near inhospitality, near unlivability, of the Namib Desert—before all that, we sit just a moment longer in the shade beneath the palms, bathed by warm but not unbearable breezes, sipping the last of a cold beer and just pretending that the world is not on fire.

We concentrate on enjoying one another's company. We concentrate on being human, and on enjoying the moment—spending time, and wasting time, even when we have none to spend, and none to waste.

Some kind of rugby championships are going on, and Dennis and Mike chat about that, with Dennis going into the vault of his college days to spin yarns about that sport, and Mike likewise traveling if not quite as far back in time, then even further in terms of distance, or worlds left behind.

We visit about the past and current legends of rhino conservation—the other legends in Namibia, whom we will not get to meet on this trip—the handful of aging and long-suffering environmental pioneers every bit as busy as Mike, and spread thin in what must feel like a seemingly endless vortex of activism. Is it too much to envision or hope for a time and place where all such battlers, such defenders, might take a rest, a time-out, for even a day, or an afternoon, in which the ever-tightening pressures of loss and diminishment are unfelt, if not nonexistent?

Sitting there with the clock ticking on his work and his life, he indulges our beginner's questions, introduces and reintroduces us to rhino conservation in Namibia, regaling us with stories about not just the labor of Blythe and Rudy Loutit, but others: the patient and gentlemanly Garth Owen-Smith, and Duncan Gilchrist, the ex-professional hunter turned conservationist, and well-known in Namibia as a raconteur extraordinaire.

Duncan Gilchrist—the raconteur—worked for the Namibian directorate of Conservation from 1983 to 1994. Originally from Scotland, he now runs a safari operation from Kamanjab, where, like Mike and SRT, he is keenly aware of the potential impact of stress on wildlife.

"The army . . . was also guilty of poaching," he writes—not just the natives in the bush. Kaokoland was a huge hunting concession for the top brass. The killing was out of control, though he laments now the unregulated presence of tourists, just out roaming. "The wildlife," he says, "is still getting hassled, chased off from water, if no longer outright killed."

Garth Owen-Smith, who has worked in the region for nearly forty years, was a critical force in helping the rhinos of Damaraland turn the corner, easing them around the knife-edge of absolute extinction. The numbers of elephant and rhino were down to puny double digits in 1977 due to poaching, and "rumors of excessive hunting of big game by high-ranking civil servants and South African Defense Force personnel had become rife," Owen-Smith has written. "To make matters worse, demand for rhino horn and ivory had increased dramatically in the Middle and Far East, and their black market value had skyrocketed."

By 1983, there were only an estimated sixty rhinos left in all of Damaraland and the associated region of Kaokoland.

As conservator at the time for the Namibia Wildlife Trust, and with help from Chris Eyre, and who was the directorate of Nature Conservation, and Gilchrist, Owen-Smith began the auxiliary game guard program. Himba and Herrero tribes selected individuals among them—knowledgeable individuals—who would patrol the region, shadowing the rhinos, and who would be paid in "staple rations and a small cash allowance," writes Owen-Smith. "By the end of 1985, it had resulted in the convictions of over 60 people for illegal hunting. The most important contribution made by the system," he adds, "was the direct involvement of the local community and conservation of their own natural resources."

In retrospect, it seems so easy: set up protective sideboards with associated incentives—reattach by whatever means possible the health of the wildlife and the landscape to the health of the human communities—and let the world continue its turning. And maybe the twenty-plus long years that have passed since Gilchrist and others were sketching these ideas—the blink of an eye, by any scale other than one human's life—are enough, finally, for this is the week of a surprise announcement: the new expansion of Etosha National Park to take in all the rest of Damaraland and Kaokoland, stretching to the coast. (Though as Mike and the others skittishly acknowledge, the devil will be in the details: You can't fence in all of northern Namibia. You almost *have* to assign tribal governance to the new park lands to make it work—and yet, will that happen where for twenty years it hasn't, and if so, will it be effective?)

It's big news, this park expansion, and yet there's no use in getting too excited yet. This too seems one of the lessons that the expatriates from wetter climes—Kent and Liverpool—have learned from the desert: *Wait and see.*

So what if a community-based conservation plan or model takes twenty or even thirty years to implement rather than six months or a year? Isn't the fit and integrity, the polish and sculpt of a thing, more important than the speed?

The more tender negotiations of human relationships—specifically, of the human heart, here in the desert wasteland—are still on Mike's mind, or have returned to his mind, for on the drive out to Rhino Camp, also known as Tent Camp, or Luxury Camp, he stops and points out another blossoming plant, a gentle little silvery gray-green shrub looking much like our sage. He breaks off a tiny sprig, sniffs it, hands it to us, there in the blazing heat, and tells us that it's a traditional sweetheart's flower: that whenever the beekeepers go up into the mountains in the spring, where they're gone for a week or so, they always come riding back on their little donkeys with bouquets of the fragrant blossom for their wives or sweethearts; that it's a regular ritual of spring, and of the beekeepers, and as such, the flower is revered.

Mike laughs, describing how the beekeepers will stop outside of the village and get spruced up before riding back in with their offers of affection.

He seems more relieved than amazed that even here, amid so much poverty—a quantum level below poverty, really—and in such a culture of strife, the gearworks of the human heart, at depth, should behave no differently than anywhere else in the world. That a man is always a man, a woman always a woman, and that even here, especially here, a flower is always a flower.

The same locust-buzzing that accompanied us all along the Hoanib is here, too, the insects seeming to be driven insane by the heat. Mike's truck is so laden with gear that I have chosen to ride up on the wooden luggage platform—the spotting platform —where the view is sublime, otherworldly, with the horizon and various mountain ranges visible in all directions, and all the herds of animals, some of which are out and about even in the full heat of the day. To stay cool, I drenched myself quickly, still fully clothed, with a cobra-shower just before we left, and now as we bounce across the road that winds seemingly without rhyme or reason through the stony plains, the breeze of our passage cools me, even chills me slightly, despite the overpowering heat. The extravagance of humanity: two gallons of water used merely to soak my clothes, my skin, my hair. I'm covered with sunblock as well, and as we motor on through the basalt plains, I am safely inured from nature—gliding atop it, looking down at it. It's not quite as good as participating in it—not quite as good as leash-walking the rhino down Park Avenue yesterday, or the staredown with Tina and her calf, Ongoody, the day before—but it's still pretty darn good, and perhaps best of all, the heart, and this mind, has no previous reference for it, no point of attachment.

Everything continues to be new, flowing in as if to fill a stupendously empty vessel. How could I have not known so much about the world—and if I am able to be so easily filled by such a simple act as a pleasant and satisfying drive in the country, then how infinite, how unimaginable, might be the rest of all that I have not seen, have not attempted to learn or see, have not even wondered about?

Herds of zebra flash and skitter before us, spilling away from our passage in spreading flows like diagrams of radiomagnetic waves pulsing from a source. They wheel and gallop as a unified organism, and their seesawing stripes and bands remind me of a school of fish. These are Hartmann's mountain zebras, with less brown shadowing between every black stripe—whereas the Burchell's zebras, which are found farther east, in a region that includes Etosha National Park—are less brilliant, sometimes more of an off-white color, with slightly different banding: thicker black bands than those of Hartmann's. (Every zebra's pattern is unique, like a human's fingerprint.)

Why? I spend a long time wrestling with the first place my mind leads me in such matters—camouflage and defense—but I can't understand how that would help out on the plains. A tiger's stripes in the jungle or a leopard's spots, I can understand: but a zebra's? Perhaps the stripes serve not to mesmerize or disrupt any rush of predator, but rather to help synchronize the spontaneous flight of the herd, when flight is called upon: the entire herd, in one sudden movement, transforming the immediate landscape around them into a pitching sea of broken puzzle pieces, fragments and slivers of black and white pitching and yawing and rolling, and with the vision of the tightly packed herd disrupted, so that now flank is not easily discernible from neck, and an incautious or befuddled attacker is as likely to get a rear hoof in the chops as a firm stranglehold on a neck.

Or perhaps the reasons are far less dramatic, though equally vital: more mechanical in nature, with the brown stripes being slightly less reflective than the pure black stripes, which might be the result of nothing more complex than physical/mechanical thermal regulation, with the mountain zebras' needs tempered and adjusted slightly from those of the zebras down in the dazzling heated saltpan of Etosha.

Whatever the reason, or reasons, it all seems proof that we're in the real world—if ever there was any question—for no dream could fit together this perfectly.

Oryx, giraffes, and springbok continue to flow before us. Behind us, the touchstone of Mike's favorite mountain, never mind the name, grows smaller, though always remains in sight, as if we are somehow tethered to the beacon of it, no matter how distant. We stop in the bottom of one steep sand wash and get out and examine fresh lion tracks that Mike has spotted. I've seen probably close to a hundred lions in my valley in Montana across the years, in all seasons, and at any hour of day or night, but here in the African summer of December, they are almost entirely nocturnal. How I would love to spy on one of the African lions, thrice the size of our Western lions, belly-slouching not through the archetypal tall grass but across these burning red rocks, this fantastic, unchanged landscape on which almost all is laid bare and made so imminently and hopelessly and mercilessly, democratically, visible.

We drive on. We have no sense of time, and perhaps there is no time; perhaps this is the cause and effect of such a landscape, and one of the reasons it has endured unchanged for so long: it devours, absorbs, or otherwise negates time.

In an instant, and then all day, it feels as if the three of us are fresh out of college again, or perhaps have only left high school, and are on a road trip, with absolutely no schedule to keep. We will get to where we are going only when we get there.

At some point in the journey, a fence—*the* fence—reappears: the rusting strands of barbed wire, and the sagging fence posts, not even ridiculous in their symbolism, but merely sad, in an era in which the nanobacteria of Ebola, smallpox, AIDS, anthrax, and God knows what else are exhaled from the wounds of an earth sometimes freshly cut, other times festering, and circling the globe then on the exhalations of trade winds, and on the breath of commerce.

Even if a flimsy barbed-wire fence *could* stop the continent's immense flow of ungulate biomass dead in its tracks, so what? We parallel the fence for a while, and watch as some springbok do indeed approach the fence and then veer away from it, though we see also that one of them is already in the "alley" between the two fences, and it crawls, writhes, wriggles beneath the hog-wire fence and then is free, across the forbidden and arbitrary border, and carrying its germs, its microbes, its essence, with it.

The impulse to set up the fence is understandable—particularly more than half a century ago—but in the here and now it is unseemly, useless, unproductive: an embarrassing artifact of how supreme our lack of imagination can be, and our fear of a loss of control.

We stop and stare at the silly little fence, and at the ocean of stone beyond.

"Whatever you do," Mike says, "don't bring up the subject of the fence with Chris or Emce."

The road veers away from the fence, finally, and something in us relaxes a bit further, and the red country before us appears a little bolder, a little cleaner and clearer, as when an optometrist clicks various lenses into your viewing frame and asks, *Is it better now, or this way?*

We descend from a slight ridge toward a sprawling level plain, and it is at the curl of this little relief, nestled against this faintest of slopes, that Rhino Camp is situated, a series of crisp and immaculate wall tents arranged beneath the sheltering shade of and within the various groves of mopane trees: an oasis of shade, if nothing else. Elegant little boardwalks travel from one dwelling to the next, and Mike has not even turned the truck off before Emce, the camp's owner, appears. She's a beautiful woman, with long, straight blond hair and sharp facial features, lively seeming, and the size of her smile on seeing Mike—it's been nearly a year—makes her even more beautiful. Perhaps it's an utterly chauvinist thing to say or even think, but her beauty seems to stand out even more fully here in the midst of such an extreme landscape, and her instant and gracious hospitality seems all the more formidable amid so many other of the classic elements of inhospitality.

So perhaps it's chauvinism, but perhaps it's also simple observation, for we have not to this point encountered anyone afield quite like her, nor will we, during the rest of our stay. She is who she is, uncommon, rare, and living where she wants to live; and the strength of hug she gives Mike speaks much not just to the depth and quality of friendship he inspires among those who know him well, but seems to carry also a subtext of the landscape itself—the splendid isolations, the splendid loneliness, with visits from friends, or from anyone, so few and far between, with the chunks of time that elapse between such visits and renewals possibly coming to seem as immense and vast as the physical spaces upon the landscape itself.

Emce directs us to our tents, the inside of which are cooled by the gentle breeze that stirs through the mesh doors. The tents are set up on wooden platforms, allowing carpets to be laid down inside, and are furnished with stunningly beautiful teak and mahogany chests of drawers and bedside nightstands and, in the curtained shower and lavatory section of the big tent, wash basins, and even a flush toilet.

Do not be seduced by this beauty, I tell myself, not even knowing why, and then I tell myself, *Oh, why not?*

I know exactly why not. The delightful feel of elegant carpet on bare feet, the solace of mesmerizing shadowy leaf dapple on the canvas walls around me, and the prospect of a hot shower after the long gritty ride atop the Land Cruiser remind me of that which I have been able to evade or compartmentalize during my Africa visit—the human suffering, poverty, and historical darkness, the deep foundation or infrastructure of blood and bone upon which this continent, more than any other, so uneasily rests. Call it white man's liberal guilt, or tourist's guilt, at seeing the truth—my truth, our truth—so starkly and suddenly reilluminated, in a manner that was at least for a while obscured from me, back when we were out in the desert, boiling our eyes out and baking our brains.

And from that rude reawakening, the realization and remembrance of how the rest of Africa lives, and of the raw deal of endless, merciless colonialism, it is no leap at all to remember, with a deep blush, the extravagance, the deep and abiding harm, of our consumptions and appetites, our hungers, habits, and addictions back home. The great feast in which we are all complicit.

I sit quietly on my crisp, clean bed while Dennis showers, and I think about art, and literature, and beauty, and about other escape avenues from these unfair and uncomfortable moral realities. It's so serene, so *elegant*. And is that right? Or if it is wrong, is not almost all else wrong?

Dennis finishes with his shower. One of Rhino Camp's stewards brings a new three-gallon bucket of steaming water and pours it into the overhead shower facility so that I too may shower.

I step into the bathroom, slip out of my sandals, shorts, and begrimed T-shirt, stand beneath the shower nozzle, twist the faucet handle, and then lean my head back and close my eyes as the warm water comes rinsing over me.

After our showers, we experience the luxury of clean clothes, and with the force of the day finally beginning to deflect, we wander over to the open dining tent, where we sit on the leather couches, looking out at the magnificent red panorama while sipping our tea, and for the thousandth or ten thousandth time, I think of how things look almost exactly as they must have at the beginning of time, and of how this is what they may well look like at the end, as well.

Emce comes over to sit and learn what Round River's interest is in the issues, and in the area, and to ask about ourselves, as we ask after her. She has run camps all over southern Africa, and when the concession for this one came available, she purchased it. She had some ideas about how it could be done, she says, and was eager to try them. And from what we have experienced in our first couple of hours, I think that she will be successful. I ask her what the challenges are, working in such a remote, or beyond remote, location, and she said that they're what you might expect. It can take a month to get a certain replacement part—for a moment I imagine how quickly the years might pass by in this manner, with each waiting-for period being like its own season: the spring of '98 equating with the radiator cap, the summer of '98 the condenser for the propane refrigerator, and so forth.

Years defined by waiting and by absence, and by the land's seeming resistance against those desires, so that surely the accruing effect on a businessperson such as Emce, or on anyone desiring or needing to maintain some sort of contact with the outside world, must be the ever-increasing suspicion, or belief, that any desires of man are extraneous to the landscape. And what a conflict of emotions such a realization must bring then to the proprietor of an establishment such as Emce's, where her business depends in large part on providing the ultimate in extraneous desires, *comfort,* to her visitors, in an environment where the concept of extraneous has perhaps never before existed.

Easier to colonize the moon, perhaps, and yet this is the opportunity that presented itself to her. This is the path—for now—that she has chosen.

There are other obvious obstacles: the animals prowling around to check things out, scent-marking their turf—lions, a little alarmingly, and elephants, likewise. And less expected was the rain. Ever the entrepreneur, Emce had initially constructed the tents so that they would be extremely mobile—portable, really, able to be towed (on big rubber wheels, as I understand it) out to whatever section of vast desert the rhinos happened to be inhabiting, or thereabouts: a tiny civilization of tents, elegant nomads, pursuing the rhinos' space.

For a number of reasons, this grand vision didn't quite pan out, with some of its various challenges revealed right from the start. The first season they were in operation, Emce invited a lot of press folks out to visit. The elegant hors d'oeuvres were laid out, the bar was stocked, the linen tablecloths spread for the magical picnic in what had always before been a golden-red dusk, with views so sublime as to stretch and expand the mind—how could Emce *not* dream of a big mobile prairie schooner of a camp?

This once certain evening, however, brought not the familiar red pixie-dust-haze-filtered view of sunset but a purple wall of storm such as one might encounter on the Gulf coast, or in New Delhi during monsoon season.

A great wind came up, howling and ripping at the tents, tearing them from their frames, pasting and wrapping them around the trees like drenched shrink-wrap, peeling up to expose the assembled guests to the storm's lash, and, as I imagine it, even beginning to set the mobile platforms in motion, with the upended tents serving as sails now, and the mobile rhino laboratories beginning to roll across the desert hardpan unguided and unpiloted, like covered or partially covered Conestogas.

"I hadn't expected the rain," Emce says.

Determined but not stubborn, Emce changed her game plan and decided to settle on one place, for better or worse. Now she has fashioned quiet elegance, quiet permanence of a sort, and while there are mornings when her guests may have to journey down the bumpy roads for an hour or more to reach whatever basin the rhinos happen to be inhabiting that day, things are probably better all around for each of the involved parties. It was a grand vision, definitely worth the dreaming, and worth the doing, even if it lasted for but an evening or even only part of an evening.

Emce excuses herself to go help with dinner preparations, which I suspect, from the look of the long dining table in the tent, is a fairly involved task. (There are other guests still afield— travelers, tourists from Germany.) Dennis and I pour another cup of tea and step down out of the tent and into the desert. I like how there are no doors involved. It's like walking through a huge plate-glass window, only there is no glass, no substance, no anything. It's like stepping into a dream. We walk out into the desert a short distance and then sit down in the wicker furniture that has been arranged beneath the shade of more giant mopanes.

How I wish I could describe the silence, and the feeling that you are suspended in another world, and another time, and that it is more real than the one you left, and yet that you cannot stay here: as if you are a visitor to another planet where the air is too thick, or too thin. You can only visit; you are a guest.

Mike joins us, freshly showered, scrubbed and glowing, looking more dashing than ever. He observes us watching the setting light out on the blood-orange stone prairie and nods, seeing that we are hypnotized, and asks, "So when will you be coming back?"

He says that it is still the same for him—that even after eleven years, there's not a day that goes by when he doesn't feel that same way, and marvel that he's able to be here. You can only visit, you are only a guest, but there is something relaxing and empowering about this knowledge. I can't quite put my finger on it—again, it has the quality of childhood—that sense of having newly arrived—but it is leavened or sweetened now with a dim understanding of how ancient the world really is.

There are no words to describe this sensation, this addiction, when you sit quietly watching that gold-red-orange light holding steady, then going away. Maybe it is like the feeling of all your old blood being drained out even as you are being infused and filled with another kind. The oxygen is different, the chemicals and elements of the universe are different, and it is as if you are standing over an entirely different history—not of the world, or of another country, which goes without saying, but of yourself. As if you never were, and might never be again.

It's intoxicating, mesmerizing. You stare into it—the orange-red-gold color of that vision, or awareness—and feel a strange gentleness, in which there is neither hope nor despair. It is—words fail—*beautiful,* and you watch it until the light is gone.

• • •

We have just gone back into the tent—candles are being lit—when Chris comes walking up from out of the desert, like some Old Testament prophet gone on safari, as if perhaps he encountered a bit of trouble and is only now able to make it back to camp some two thousand years later. It's sloppy writing to say he's a bear of a man—for one thing, there are no bears in Africa, or even in Scotland, where he's from—but he's got a force, a radiant presence, that comes partly though not wholly from his hulk. He's also only got one arm, something we'd heard about in advance, from someone back in Palmwag—a crocodile bit it off, and a lucky thing that the crocodile didn't get all of Chris—and though we notice the asymmetry, it does not in any way diminish Chris or his presence.

Chris is coming to greet all of us, and it strikes me that Chris and Mike are almost like rhinos themselves, occupying such vast expanses of territory that their paths might cross only every several months. In this instance, a full year has gone by.

Chris, having already guided some tourists earlier in the day, had gone out for a hike once the heat of the day was beginning to bend, and the three of us make note of how that seems to be the mark of a happily employed man or woman: doing in their off-time that which they have just finished doing in their on-time. Chris agrees, turning to look back out at the dimming red richness of nothing—but so *much* of it—and says, "I never get tired of it. Each day it's more beautiful to me." The last of that red light is draining away, the violet onrush of darkness is sliding in again quickly, but it does not matter: the next day will bring every bit as much richness as has this one, as will the one after that, and the one after that.

We sit in the tent and listen as the two friends catch up on their absence. Both men are undecided about the impact of the week's news about the expansion of the park, through Chris, despite not being a big fan of government regulation and intervention, says, even if sadly, that he thinks the park "is the only answer"—though he agrees fiercely with Mike that the new expanded park country must be managed by local cooperatives.

Chris and Mike commiserate over a recent article in Europe that profiled one of the dry river washes, with maps and other how-to-get-there information, with the result being a horde, an influx of uninformed and uneducated tourists coming down from Europe and running amok amid the elephants. The elephants have already killed some tourists. "Enough is enough," Chris says, and he is not siding with the hordes of tourists.

The "uncontrolled, unregulated traffic" is going to kill the region's wildness, he says; and again, I am struck by how he could be speaking of northern Montana, or southern Utah. We in the West are already ankle-deep in the boom, while Chris, like a prophet, is still only staring out at the horizon, feeling the sonic rumble of it rolling his way. Unlike us in the West, he's ahead of it— just barely.

About the park, and his begrudging realization of its necessary expansion, he insists, "They [the government] have got to look at partnerships with some of the conservancies to get greater buy-in from the communities." He's speaking of the village- and watershed-conservation districts that folks like Garth Owen-Smith and others have been laboring to help set up these last many years.

It all might just possibly come together, just barely in time. So much work—so many small projects and big dreams, for so many years, and with the issue coming down almost literally to a question of hours, with the park expansion being announced just this week. Can the local conservancies handle it? Is the beginning of an infrastructure for sustainable economic development in place? Rhino Camp certainly is, stuck out here with its flag of the future planted like that of some lunar colonist. What if folks like Garth Owen-Smith, as well as Blythe and Rudy and Chris and Mike and Emce and the small handfuls of others, had slacked off for an extra day or two, a week, a month?

What if they had not gotten one certain pissant grant or another, here or there? Would the wave have passed them by, washing away this opportunity, or rather, making it into a negative, absent thing—a thing that never appeared?

And what path would that have left available to the rhino, then?

On this continent too, then, it is the activist's dilemma. No solution is entirely golden—always, the endgame seems to be one of choosing a lessened rate of diminishment—and even as one must conspire to protect oneself, to keep from burning out under the relentless pace and hence being rendered ineffective, so too must one also somehow learn to acknowledge and deal with that relentless pace.

I can't believe Chris just went off for a walk like that, at day's end. No pistol, no spear, no nothing. A lion could have gobbled him up as if he were a khaki-clad hors d'oeuvre. But maybe, having already been chewed on, Chris has other concerns, and is addressing other, more imminent and internal dangers. The danger and violence of a cluttered mind.

The other guests arrive, riding back into camp in the big canvas-topped transport vehicle that Rhino Camp takes out into the desert—so much easier than a rolling campsite—piloted by the affable and intensely knowledgeable guide Felix. They're cutting it close—dinner will be served soon—but are excited, having worked long and hard for a sighting, covering much ground, but finding one finally, and observing the animal, unobserved themselves, at distance.

"It has never happened that we have not found a rhino," says Felix. Some days are harder than others, Chris says, particularly at this time of year, when, following the brief rains, "the leaves are coming out on the slopes, which leads the rhinos away from the beaten paths."

The rains are like a pulse, a contraction and expansion, I think, and for a moment my mind jumbles things together (Chris has handed me a gin and tonic), and the pulse of a year, with rhinos ebbing and flowing, seems little different from the pulse of the millennia, in which the rhinos recede from their former range, their former colonies, but then—with the help of but a tiny handful of people—expand back into those lands.

The other guests join us, steam-scrubbed and glowing, and dinner is served—steaks of gemsbok loin, and a creamy potato gratin, fresh salads, and newly baked bread—the courses arriving with an alacrity that suggests to me the various dishes have been almost ready for quite some time, with Emce and the chefs watching and waiting, biding their time, as one after another guest showers, the chefs trying to hold off on the final pieces until the guests, the quarry, emerge from their tents and begin striding toward the tent. The chefs hovering over their skillets like ospreys, or other hawks or harriers, treading the air, waiting, with the blue flames turned low.

All four guests are from Germany—two older couples, of whom only one, Manfred—clearly their leader—speaks English. Manfred, the executive of an international company whose function I can't quite understand or fathom—something to do with imports and exports—is, I suppose, one of those folks for whom money is no object.

The long table in the tent is beautiful in the candlelight. The food glistens, the guests are conversing in German, English, and Afrikaaner, and I'm keenly aware of how fortunate I am to be here, and yet I am also nearly crushed with longing, with good old-fashioned homesickness, wishing that if my family could not be here with me, that I could then be with them—playing catch with the girls, or helping them with homework, or just hanging out, listening to them putter around in the house, and in their lives. I am grateful, in missing them, to not have an appetite for more of the world than I already do. To know my appetite, and yet to hold it in so fiercely.

At the table, Dennis and Chris, bear man and rhino man, are hitting it off famously, the two of them trading stories about grizzlies and rhinos, about Alaska and Africa.

They're talking a little about their injuries, too—not so much about the circumstances of them ("I was hit by an airplane, up in British Columbia," Dennis says simply), but instead in great detail about the long and challenging physical rehabilitation. And in what is not said—but in the familiarity with which they bond— I can see that they share also, in ways the rest of us cannot know, a mutual understanding of how hard the psychological rehabilitation was for each of them.

No sooner has dinner been served than Dennis peels back his shirt to reveal the purple and pink twisted weld of the propeller's bite, explaining how they cut out muscle from his chest and sewed it over the top of his shoulder.

Chris announces that they had to carve up his buttocks to help make a graft over the wound of his missing left arm.

"So you were able to get some movement back," Chris says approvingly, watching Dennis rotate his upper arm stiffly while curling it upward also: a movement performed perhaps thousands of times each day, unthinkingly, by any of the rest of us.

"Yes," says Dennis, staring down at his arm with—what? pride? wonder?—no resentment, no bittersweetness or anger, finally, only the moment. "Yeah, they were able to fix it where I could still turn it."

"Ahh, that's nice," Chris says, his rich, calming brogue acknowledging that he knows, at least, of all the thousands of hours of therapy and rehab, the despair and fury, the rage and surrender, the resurrection of hope—the whole bloodiness. "That's nice," he says again, examining his Guinness, and he means it, and examines his glass as if the victory he is savoring is his own; and if there was ever any question about whether I liked him, or how much, it is settled there, for sure.

The synchronicity, if you will, is wonderful and alarming, but it is not so much the two big men losing their opposing arms in their various physical struggles, continents away. It is the similar tale of survival and recovery: the jumbled reassortment of body parts that resulted in saving them and restoring some if not all previous function to them, allowing them to continue on, reduced in some ways though strengthened in others.

The story of Dennis or of Chris is the story of the African continent, or the story of Pangaea, and Gondwanaland—one cleft of landmass being torn off from the first earth and being thrust, or drifting, half a world away, to form a new Eden. As if every story is the same, whether played out across the eons, or in a single afternoon, a single evening. It is even the story of Mike, lifted from England and applied, like a bandage, over the wound of Namibia, and it is the story of the rhinos themselves, some of which—having now begun their own recovery—will be lifted up and reapplied, reassembled, elsewhere into their former range.

And I am not unaware, either, that what I am seeing is eerily the story of the hidden or silent violence to the north, in Rwanda and Darfur and the Congo—the mass amputations and genocides, limbs being lopped as if but tall grass before a merciless scythe. As if the world has already decided that in this place and time, limbs must be lost, whether peaceably or unpeaceably. As if such a world might have somehow decided, in like manner, the physical shape, physical manifestations, of all the rest of the world: that fish-catching eagles and ospreys should have snowy white heads, whether off the coast of Kamchatka or in Utah; that the pronghorn of Wyoming and New Mexico should have the same bands and tones and colors as the springbok of Namibia.

There can be experimentation and variance, in such a story—but still, so often, it seems that there is but one current moving beneath us all.

How the world needs adventurers—bighearted, goodhearted adventurers. I used to believe that the percentage of them in any given population, at any given time, would always be constant—that such adventurers existed among us in a steady percentage, as if perhaps having some rare genetic marker that disposed them to a life of such daring and exuberance.

But what it seems like to me now is that despite a general increase in the overall population, such individuals are getting even rarer; that that tendency is being washed out and deleted, is blinking out. And whether this is disappearing deep within us, waiting to be summoned, perhaps, by the return of wilder and more complete landscapes, I cannot guess; I know only that each year I encounter fewer and fewer such rhino and bear men, rhino and bear women. They are still out there, but the world—our new world—more than ever, seems to be conspiring against them.

Before he was a rhino guide, Chris was a big game guide and hunter; and before that, he was in the army, a member of the South African Defense Force (SADF), where he did see some of the poaching that's been reported from that time.

"Our commanding officers would be hunting to feed the troops. Elephant poaching—that did happen—but what I'm trying to find out now is, was it an opportunistic thing?" Chris has been going around since that time, interviewing other ex-soldiers, trying to find firm proof one way or the other, and while I sense that he hopes there is no proof—that the government didn't have an organized ring to fund the war off of ivory and rhino horn—if he finds out that his war, and his government, did operate under such a conspiracy, he will report the facts. It's awful enough that such goings-on funded the war in any way at all—one terrible thing promulgating another terrible thing. But for that to have been part of a desired and conceived plan would be much worse.

"The Himba would be asked to bring in rhino horns to SADF," he says. "Each Himba man was issued two hundred rounds and a .303 rifle. There was an underlying current that this was the reason"—not for defense against any dreaded Communist hordes trickling down through the impassable desert, then, but for the procurement of horn and ivory.

Chris remembers hearing about a herd of elephants that got pinned up against a mountain and were allegedly executed, slaughtered by the army, for their tusks. "That's what we must find out," he says—if such things, where they happened, were the result of a patrol getting bored because there was no action, or if there was a sinister ring that channeled the rhino horn and ivory for personal gain and to pay for the war effort.

After the war, Chris got a job leading an anti-poaching unit, which he led all over Namibia, chasing down poachers and, where possible, capturing them alive to bring back for trial, though there were the invariable shootouts with the poachers, particularly when the poachers were caught red-handed. Too often, they were the kind of men who found themselves desperately disinclined to avoid the possibility of captivity at any cost.

Chris tells us of one engagement in which a particularly large and aggressive gang of poachers was moving north through Namibia at a lightning pace, killing so rapidly that the anti-poaching unit's trackers couldn't keep up with them. Always, they were a day behind, following the tracks to the site of an already poached elephant or rhino, the horns or tusks sawed off, and the poachers—with many porters—on the move again. The poachers still poaching, but hurrying, perhaps at a run, as if knowing that Chris's patrol was still after them, pushing them north.

Chris realized that the only way they were going to catch them was to split up and have one group try to get out ahead of the poachers, which is what he did, taking a smaller group and climbing up on a ridge to do so.

The trackers below him kept following the poachers' bloody trail, but on the ridge up above, Chris was able to see the ravens and vultures that were attending to each kill; the birds, as I understand it, beginning to associate with the poachers: learning that where the poachers went, there would be meat, so that Chris's group could run toward the vultures, could follow them straight to the poachers, which they caught just this side of the border, just before they escaped his jurisdiction. They were able to capture the poachers alive and send them back for their trial and their relatively minimal fine.

Long after Chris has wandered off to bed—following a last nightcap and an impressive five-minute dramatic monologue, perfectly rendered, of Robert Service's poem "The Cremation of Sam McGee"—Dennis and I remain, watching the mopane burn down to mounds of ash.

And when we finally rise and return to our tent, the net of stars enveloping us with a different kind of electricity than in North America—cleaner, lighter, leaner, somehow—we find that our beds have been turned back and that, in addition to the elegant little chocolate truffle that has been placed on each perfect pillow, there is a hot water bottle nestled beneath the sheets, in the center of the mattress.

A naïve Yaak hermit, I've never even heard of such a thing, and neither has Dennis, so that when we pull our covers back and spy the pinkish, natal-looking thing, plump and elongated, infant-size, lying there, we yell simultaneously, then touch the odd item cautiously, and then pick it up. The heft of it is fluid, like a big fish, but warm.

The whole bed is slightly warm from that bottle, and we each place the little infant back under the sheets and then slide beneath our covers, in a warm bed in a cold tent under a violet to black starry sky. This is Manfred's world, not my world. I have to admit, the little water bottle is a hell of an idea, but I think people like one-armed Chris, and almost-one-armed Dennis, and Mike, doing their community work—Mike, in particular, careening about Africa in his Land Cruiser, with its "A Good Heart Never Fails" bumper sticker, dedicating his life to the empowerment of the cultures that have staked their lives on this intersection of landscape and self—are like magicians.

We all know such people, from and in the different walks of life. They are here for a purpose—sometimes realizing it, other times appearing not to. They are sometimes charmed, other times not. They glide through life accomplishing great good, more than would seem possible: sometimes through muscular dint of force or will, other times as if through the workings of some strange and unseeable catalyst.

And once again, I consider the image of a small boy in a Liverpool suburb walking a few houses down to stare through the hedges at the sight of a pen of rhinos being fed sweet green hay and dusky *Euphorbia* fronds just flown in from the desert so far away: the *Euphorbia* so fresh that the latex sap was still oozing. And something happening in the boy, and in the world.

The thing I hate worst about writing—about poor writing—is the use of abstractions, and unearned similes, and unearned metaphors. And yet in trying to describe what daylight is like in the red desert of Namibia, with the sun rising over those low hills and illuminating the flat-pan stretches, and giving cooling shadows to the scattered lone *Euphorbia* and mopane—the desert and the rising sun both images of fire, and yet with the red world so still and so cool, so peaceful, even as such fire rolls toward the viewer—I find myself seduced nonetheless by the greed, almost a gluttony, of abstraction. Because the experience, the sight, is so encompassing, you want to lay broad claim—you do not want to leave out any subtlety, any emotion, any comparison.

It is *beautiful*, you might say, or it is *peaceful*, or, more likely, it is *so awesome*. When I think of it, my own abstraction pulls me to the description that it is like childhood, but then I founder, unable to explain why, only that it is intensely so. Each morning reminds me of what mornings were like as a child (those memories have nothing to do with the external landscape: the childhood of which I am speaking occurred in the shadowy, forested suburbs of Houston, alongside a slow-moving muddy bayou, amid magnolias, Spanish moss, big pines and oaks, on a street named Shady River). Each morning here reminds me of how the day, and the world, felt to me each morning as a child—and it surprises me, almost overwhelms yet also steadies me, to remember how much strength and confidence and imagination and faith I possessed back then; or rather, did not possess, but stood amid, perceived, witnessed, observed.

I could see it, then—it was everywhere—and then after a long time, I did not see it so much.

But here in Namibia, it is back. It comes rolling in every morning—the joyful beauty of the new world, like a child's world, insistent and full of promise, here on one of the oldest or most unchanged patches of earth in the world. The turtledoves begin to call, in that rising light, and sometimes the mornings call to you to hurry and take a walk in the cool air and the shimmering gold-red light, as the haze of dust motes begins to spin and rise, and other mornings the world and its promises call to you to merely sit quietly, suspended in time, and to watch, but not yet step out into the world.

I can do that. Some mornings, I go for walks in that first light —deliciously aware that there are lions out on this landscape, and that I would do well not to encounter one—while other mornings I sit quietly and sip coffee and move barely a muscle, simply absorbing the day, as are the rocks and stones around me: waiting, as if to come back to life.

Here in the luxury of Rhino Camp at day's first light, I am

sitting by the sweet-scented smoldering remains of last night's campfire, the burned-out husk of a mopane stump still exhaling tendrils of dense blue smoke, and listening to the doves, when I hear a new sound — new to me, at least — that seems remarkably like the squeak or croak of spring peepers.

Nothing about this landscape would surprise me — or rather, nothing about it would seem beyond the expanded range of possibility. If there can be aquatic turtles in this desert, then can there not also be spring peepers?

Toads, more likely, though — some fabulous variety adapted to the fantastic boom-and-bust pulses of heat and drought, emerging in the brief windows of thunderstorms, biding their time beneath the stones and beneath the soil, waiting for decades, perhaps, for just the right moment, just the right confluence of events, to emerge and make a gamble, an expenditure of all their various resources and instinct and, if I dare say it, *hope*.

Listening to their calls, so near and clear, I determine that I must see one, and so I rise from my camp chair and walk quietly out into the desert toward a spot that I perceive to be one of the sources of the calling.

I stalk the sound, hunkered low and crouched over, as stealthy as a waterfowler sneaking up on ducks quietly gabbling on the other side of a levee. I'm tiptoeing across the desert, approaching the sound so cautiously, and am frustrated when, as I draw near to the rock from beneath which I think the sound is emanating, the calling stops.

Now the song comes from another location in the desert, from beneath another rock — a nation of toads beneath hundreds of rocks, communicating with one another, luring me off the trail whenever I draw too near to one of their number — and I pivot and begin sneaking toward that unseen caller, only to have that one also fall silent as I draw nearer.

For fifteen or twenty minutes, in that rising sun, I crouch and sneak across the stony hardpan, as if following some maddened choreography of evasion. And although the calling locations are multiple, there is never more than one caller at a time, so that it seems to me that all the frogs are taunting me with their calculated silences, before calling again, after I have passed by.

Exasperated, I straighten up and stretch my back and scan the desert, which is now golden rather than red. The day is getting away already, the temperature is beginning to click upward—the dampness of the day's first perspiration trickles down my lower back—and I see that from the guides' tent, Felix, dressed in his khaki shorts, roomy khaki shirt, and broad-brimmed hat, is sipping coffee and watching me with equal parts concern and puzzlement. He's trying not to stare, and yet . . .

I wander over to where he's standing and explain to him about the toads. His brow crinkles and he actually takes a step back from me. I suppose he sees all kinds of people out here, and I can see him trying to evaluate me for the coming day's journey, assessing the problematic nature of me riding in the midst of his clients, his responsibilities. *The damned Americans and their recreational drug habits,* he might as well be thinking. Does he think I was out there taking my morning constitutional hit of acid?

Thinking back to my strange and intent dance steps, I do not see how he could believe otherwise.

"Did you see any toads?" he asks cautiously, politely.

"No," I tell him, "every time I got close, they got quiet." And though the toads have fallen largely silent, as I had feared they would, as the day warms (why did they not call in the night? I wonder), the call does begin once more.

Felix shakes his head. "I've never seen any frogs or toads in the desert," he says. "I don't think they're out there."

"*That* sound," I say, gesticulating toward the horizon. "Is it coming from beneath the ground? Isn't that a frog, or a toad?"

Felix gives me the *Is this American tripping?* stare once more, then breaks into a broad grin. "Those are cory bustards," he says gently, and points toward a lone mopane tree that stands a good two hundred or more yards away—easily three or four times beyond where I had been turning over stones, searching for toads.

In the lozenge of shade beneath the tree—at the edges of that shade—there is a small flock of bustards, looking at this distance like guineas, scuttling back and forth and pecking at insects beneath the dry curls of last year's leaves. At first I think Felix is teasing me—how could any sound carry that far with such crystalline frog-croaking clarity, for me to believe it was coming from only a few feet away?—but then I remember quickly, *Africa,* and I say something lame, like "Gee, they sure sound like toads."

As delighted by my being fooled as he is relieved that his day's duty will not include the ferrying of a troublemaker, Felix hands me the heavy-duty industrial-desert-strength guide's binoculars hanging on his chest, and through oculars with roughly the magnifying power of twin telescopes, the feathers of each bustard leap into variegated clarity. Their eyes gleam and burn with the day, looking back at me across that distance, probably with equal clarity.

Were they scuttling about, changing positions, slightly threatened, as I frog-stalked 150 yards shy of their morning maneuvers? And what kind of bird makes a chirruping, croaking sound like a toad anyway, and why?

I mumble some excuse about needing to prepare for the day
—he nods, and gives me the grace of glancing at his wristwatch
as if to agree that yes, that would be a good idea—and as I hurry
down the boardwalk, still feeling the heat of my blush in the cool
of morning, I know without a doubt that I have become part
of future camp lore, the kind of story the folks in camp will tell
to other newcomers such as myself, after I am gone. Toadhunter.
Frog-Searcher. *He was out there before daylight,* they'll tell their
dinner guests. *It was the sound of rocks being overturned that awakened
me. It was a hell of a racket. I looked out my tent and saw him out there
in the middle of the desert on his hands and knees, digging, looking for
toads . . .*

We dine in luxuriance, and again, I feel the American's creep of
shame. It's good to see Mike pampered, at least.

Felix, commander of the visitors, gathers us into the big military convoy vehicle that's been outfitted to become a luxury tour
bus. The top is cut off to enhance viewing, with a high frame
welded above the tour bus to provide a canvas canopy, shade
against a sun and a sky that would crack our noggins like melons,
would otherwise fry or bake the tender fecund swamps of our
brains.

There's no air conditioning, save for the heated dog-breath
furnace blast of tickling, chafing, broiling wind caused by our
own movement, but the seats are super padded, and the bus is
well stocked with ice chests, and we move out into the brightness
of morning in a military vehicle, but unencumbered here by any
realities of land mines or enemy fire.

It seems fragile, this newer reality of the moment, and almost
unnatural, so unaccustomed are we, psychically and historically, to
a landscape so vast, sere, and uninhabited, unclaimed, and seem-

ingly uncontested: such a vast pool of potential, and such a vast pool of peace.

I remember reading something by the writer and historian Gary Ferguson, in his book *Shouting at the Sky*, about traveling with troubled adolescents who were enrolled in wilderness therapy courses. When in the shadowy forests, a lot of the hard-case kids wouldn't open up and "be real" with themselves or their counselors. In the deep woods, hiding was still a psychological option. But when those same youths moved down into the desert, they would "get real" within a matter of only two or three days, opening up—cracking open, is how I imagine it—and crying out, the rage and all else issuing from them as if from some volcanic vent beneath them, with nowhere else to go now, nowhere else to hide.

Here, there are no angry youth—only space, peace, and the guardian rhinos, attended now, occasionally, by their guardian observers, with the rhinos themselves rarely even aware of those guardians, or that any force in the world is observing them.

Felix misses nothing. He's as good a game spotter—a hunter— as I've ever seen, and an acute naturalist. A tiny wisp of red smoke rising from beneath a far-off mopane—a thread I cannot even see bare-eyed—is revealed, through lifted binoculars, to be a flock of guineas scratching in the leaves.

We drift as if in a sailboat on gentle swells far out at sea, over one slight rise and down the next. Again, the mystery prints of unseen passers-through stipple the red sands—elephant, lion, giraffe—while the zebras, oryx, springbok, and kudu flow back and forth before us, washing up and then down the same hills, also like waves. Felix can tell the sex as well as age of each of the distant ungulates, based on the way they're standing, or the way they're walking, or the way they're interacting with the herd, and there is rarely a span of time in which he is not showing us something: not just the tour guide identification or naming of things, but explaining to us the process and function of each named thing and the way it relates to other things. Gradually, as has been happening steadily throughout the trip, I come to understand that this world is not just intense sunlight and rhinos, and red stone, red sand, but that there is another galaxy of life, even if largely hidden, hiding in every wedge, every niche, every opportunity of the imagination. No toads out here, but almost everything else.

Under optimal conditions, Felix tells us, the rhinos can go days without water.

Not even a *machine* can go weeks out here. How different, really, is a rhino from a rock, a boulder, able also to go weeks without water? Are they not then simply like rocks inspirited with the brief condition of life, the spark and fire of life?

Out on the desert plains, the *Euphorbia* in the rising sun looks strangely pastoral, like round bales of hay—though from what pastures such gleanings might have been gathered seems a mystery, as inexplicable as the hoofprints of the giant night travelers before us, unseen.

I look around at the blank tableau, or nearly blank tableau, wondering what Felix might say next to fill the space, the silence, and do not have long to wait. There is nothing before us, only sun and sky and stone—not even a lone oryx—but he slows the truck, stops, and gestures to one of the tracks before us.

"You might notice that the feet of desert-dwelling elephants are larger," he says. "It's because they walk on much softer ground." Like the broad feet of snowshoe hares and lynx, the feet of the desert elephants have adapted to the dunes.

What, at this late date, is not by now perfect about the world?

The obvious answer that springs to mind is *us:* we the newcomers—and again I am forced by that evidence, the daily proof of our imperfection, visible anywhere on the globe, to reconsider our old stories, old models or beliefs, old perspectives, and wonder if in the rush of our newness we perhaps got them wrong somehow, fumbling even our scale of time. What if we got it all wrong, and the Judeo-Christian creation story in Genesis and all the world's creation stories were not divine history revealed but prophecy of a world or worlds still to come?

What if such stories are like a manual of sorts, a code or set of operating instructions, with the first and then all subsequent sculpting of rib and clay and mud still in progress, with any of the story's sentences never to be read with the finality of a sentence's end, but always as sentences and stories still in progress?

If that is the riverine way the world is, then what are we to make of those smatterings of landscapes in which such flow, such fluidity and flexibility, does *not* seem to hold true, but in which time—after almost all other things appear to have been sculpted —has finally paused, maybe even stopped? And what do we make of the creatures that we might occasionally find in such places that appear, for all intents and purposes, to have likewise paused or even stopped, standing sentinel now over what might well enough be called perfection?

And what does it say to our newness, and the fire, the not-yet-cooled turmoil in us, that we as a species, should try to kill such sentinels?

We glide through the heat, the big army truck rumbling, yet with the padded seats pitching us around as if we're on a bed of pillows. We can feel the stone wall of heat just beyond the shade of our canopy, but there's ice water, iced tea, iced Coca-Cola.

"I wonder how the party went," Mike says. He does not sound wistful, and no one says anything, we just ride on into the heat, gliding, another day deeper into the search for rhinos.

The radio crackles, information is transmitted, and Felix wheels the big bus around in a slow one-eighty, and we begin traveling toward the sun, and into another valley, where, in the distance, we see the glint of the jeep. The trackers are waiting for us, staring down into yet another basin, where they say a rhino and her calf have just walked around the bend of a mountain, about a mile distant.

There's no good way for the jeep and tour bus to go down the bluff on which we're standing sentinel, and so the decision is made to loop around to the far side of that same mountain, park the vehicles, proceed up a dry wash, and hope to intercept the cow and calf on what Felix hopes remains their trajectory. He glances at us, assesses our hardiness, factoring in all the various distances involved, and rates of travel for jeep, bus, rhino, and tourist across the various routes, and then says that if we hurry, we might make it.

And that's how it works out. We hurry up the dry wash, stumbling through a garden of blood-red basalt, in which the teeth of thousands of magnificent quartz crystals blossom—as if we are strolling a wandering path in a garden in which snow-white lilies bloom—and once again, we find that we are racing the rising heat of the day, with the hammer and anvil of sun and stone conspiring to send the rhinos, and all other creatures, quickly to the shade of a day-bed to wait out the middle of the highest, hottest peak.

We come to the mouth of the wash and spy them, another mother and her calf, just as Felix had envisioned, moving away from us.

Felix has a little spotting scope with him, a rangefinder, and he asks me to guess how far away they are.

I once shot an antelope on the Montana prairie at a distance of 323 yards: not a very long shot for some, but far for me. (I wouldn't have even attempted the shot, but believed the animal to be much closer, a little less than half that distance.)

Remembering that antelope, I totally make up a number. "Oh, about three hundred and twenty-five yards," I say. Believing, again, that the rhinos are really much closer.

Felix looks at me suspiciously. "It's three hundred and twenty-four yards," he says, and I make a *tsk*ing sound, then tell him, "Well, your machine is off; you'd better get it recalibrated."

And again, the race is on. We can see the tree toward which the rhino and her calf are headed, about a quarter of a mile away —the only tree, the only shade—a lone oryx rests already beneath it, silhouetted and motionless, like an ornate chess piece. The wind is in our face, so that we're able to hurry along behind the rhinos, trying to get into better viewing position before they collapse into that one ellipse of shade that surely they cannot see, with their dim eyesight, but that the mother must know is there, either from memory and experience or maybe by scent—the cooled patch of distant earth exuding a different odor, perhaps as pungent and attractive to her powerful sense of smell as a busted-open crate of strawberries, or the odor of a marsh, an oasis, shimmering in the sun.

We are unable to close the gap—the rhinos are traveling too resolutely, and Felix insists that we travel slowly, carefully, guarding his resource, *the rhinos,* extra cautiously in the midday broil —and from a distance of about three hundred yards, we watch as the oryx vacates the shade before the rhinos' approach, trots out into the unbearable sun, and then just stands there, ceding his territory.

One must be careful about anthropomorphizing any animal, and especially in a new country, and with a species totally unfamiliar to the observer; but I have to say, it sure looks like sulking to me: and though there is no one to appeal to, save the sun and ourselves, the oryx continues to stand there, baking and watching.

The rhinos circle the mopane, dark shapes now moving around in that tiny pool of shade, as if swimming laps in it. We crouch among the crystal fields and stare through our binoculars and through the rising wavers of stone heat, and we fall once more quickly and deeply into that reflective state of peace that accompanies the simple or not so simple act of gazing upon such improbable creatures. I think that we could sit there, despite our own lack of shade, and watch them all day long.

After one brief turn around the mopane, the mother plops down in the dust and lays her head against the ground in such a way as to be sure her head, and the massive horns, receive the maximum amount and highest quality of shade, and I think again what an enormous heat sink the horns must be, absorbing and then dumping excessive solar radiation.

The calf is fidgety—there is not quite enough shade for both of them—and he appears also to not be ready for his naptime, desiring instead to stay up and play. He lowers his head and shoves against his mother's flank, and whether to try to get her to stand up and play or to bulldoze her a short distance aside in order to obtain his own fullness of shade, we cannot be sure. She does not budge, and the day, and time, feel oddly frozen, with the oryx just standing there, with nothing to do, and we tourists just sitting there out in the broad sun, and the mother napping.

Among us all, only Felix is less than rapturous. "I think the mother is very tired now," he says, and from his concern, I can tell that if there was anything he could do for them, anything to make their way in the world easier or stronger, he would. But there is nothing; they are as they have always been, on their own, and the best gift we can give them is to do no harm, to stay back and allow them to continue their daily and ceaseless negotiations, perfectly fitted with a relentlessly harsh and unchanging landscape.

And again, in the heated mush of daydream, it occurs to me that to some fundamentalist interpretations, this near-waterless, broiling lonely land could be viewed as an archetype of hell, even as to others — Chris, Mike, Felix among them — it could be viewed as an archetype of heaven.

Either way, there are aspects of eternity everywhere we look here: hints, clues, and scraps of evidence, that are almost as real and physical as the stones themselves scattered across the desert hardpan.

Felix remarks casually that oryx calves are born with horns, ready to fight, and to defend themselves right from the beginning.

Who dreamed the farthest and furthest corners of this dream? How much of it was known beforehand, and how much has been built and fitted according to opportunity? The answer seems doubly frustrating, ultimately unanswerable, despite being closer than perhaps anywhere else in the world. As if it is now only 275 yards distant.

The calf gives up on his hopes for play and snuggles in to the only irregular shade that is available to him. He rests his head against his mother's flank and closes his eyes and goes to sleep; and out on the desert, time, as well as the world, stirs, and begins moving again, though carefully, and respectfully, as if—like us—afraid of awakening and disrupting the rhinos, or changing them in even the slightest.

And what else might that be, then, but the definition of love? The world, and its maker, loved, and loves, the rhinos and the Namib: loves them.

The wind stirs as time, authorized now to do so, slides forward with greater alacrity. The same wind that bathes the rhinos sweeps across us a moment later. It bathes us, too, with only the smallest amount of distance separating us. And are we then similarly loved, or might we one day be?

If *love* is not the word for it—this pride of craft that exists in the fitted world—then what is the word?

They are so close: less than three hundred yards away. Felix tells us how some tribes' children, when bored, will play a game in which each child takes turns sneaking up to a sleeping rhino and placing a small stone on its back —approaching so lightly and stealthily from downwind as to avoid awakening it—and then retreating, with the next child's challenge then to similarly place a slightly larger pebble or stone: and so on, until the rhino awakens and throws off his or her coat of stone, scattering rubble in all directions as if having just emerged from beneath the rocky skin of the earth.

It's a way to pass the time, Felix explains. It's something to do while the world is sleeping, and—though he does not say this—a way, perhaps, to cross or bridge the fifty-five million years that are separated now by only that last short distance.

We rise, however, and make our way back toward the tour bus, retreating rather than advancing. The distance widens to a thousand yards, and then two thousand, and then a mile, and still farther; and I know with greater certainty than ever how much the world needs places where we will not build roads or dams, or mines or villages, and that the Namib is one of those places, and that further, Mike's work, and the others', is surely blessed for his attempts to nonetheless try to find a way that accommodates these seemingly incompatible needs—the need for local communities to survive, and the need to not destroy or even alter a made and perfect thing.

He's young. He just might be able to do it. He is a force of the world, like the wind itself. He fits the moment—fits the present—and, I hope, fits the future.

Back in Rhino Camp, we say our goodbyes and thank-yous. Dennis and I will be traveling on to Etosha National Park—far and away the world's largest fenced-in wildlife preserve—and Mike will be traveling back to Palmwag, where he'll finish up three or four days' worth of paperwork, then travel down to a place on the coast called World's End, for a couple of weeks of vacation, between assignments. He'll ride his mountain bike, and go surfing in the cold South Atlantic. He'll visit friends, will rent a little cottage in town during that time, will come in from the desert hermitage, and practice being a social creature again.

We embrace before parting. "Come back soon," he says.

"Thank you," we tell him, and the others, again. "We'll see you again soon," we tell him.

And driving on then, into the wide brilliance of the day, Dennis and I both feel that opening-up feeling, and of the world's newness, so much so that we comment on it, and on how rare it is for men in their late forties and early fifties to make strong new friends, so late and far into the game.

There's no explaining it. Either there are larger plans and destinies, or we just got lucky. Either way, we have a new friend, loyal and good and true, and slightly cornball, idealistic, earnest, and strong. A young man with a good heart, and one who is willing to spend it, to spend it all.

We drift on, through and across the desert. We follow the ridiculously porous and ridiculously imagined ancient rinderpest fence—folly!—and then veer back out into the rank vastness of desert. On a nearby hillside, we see four giant kudu jammed together beneath a single shepherd's tree, all four of them positioned Picasso-like in order to utilize every available slant of shade, and holding motionless, paused there while the summer sun pauses too before releasing them, the shadowy bars of shade melting away as the sun rises ever higher. The vanishing shade releases the four puzzle-pieced kudus back into the heat, and they are propelled, whole once more, but suffering, or challenged, into the current of time, stepping gingerly across the moonscape and into the future, their ankles wavering in that heat as if wading a silvery river, or as if even the kudu themselves are melting, as are the stones around them, once molten and then hardened but now melting again. As if all of it, the living and the unliving, was never anything more than an idea, a plan or scheme, a dream.

Epilogue

Etosha National Park is a different kind of heat, to a connoisseur of sun: saltier, more humid. A heat-warped two-lane blacktop road wobbles into the park's southwestern entrance, closer to the Botswana border, with the road passing through landscape quite different from the stony austerity of Damaraland.

Clotted brush and thorn-weave crowd the road, and it is a rude shock to see barbed-wire fences, driveways, gates, and even an occasional billboard: as if we are in rural south Texas — as if Namibia will not hold still for our vision of landscape, and certainly will not hold still for our archetypal primitive vision of culture.

At the park's entrance, there are only a couple of vendors selling brightly colored wood and straw crafts — mobiles, carved wildlife art, bright blankets — and then we pass through the stone wall gates, which remind me strangely of Yellowstone's Gardiner entrance, and then we are into the official park, where we take the first turnout, travel down a pale gravel road to a cul-de-sac, to an open area of plains, where a herd of two dozen zebra are walking in front of us not thirty yards away.

We stop the truck and watch them through the dust- and bug-smeared windshield in the late-day sun. You can't get out of your car in Etosha, not for even a second; it's a park rule, strictly upheld, designed to keep the lions from eating people.

The zebras are beautiful, muscular and vivid and wholly un-concerned by our presence, unlike those out in the wilds of Damaraland; but parked there at the end of the cul-de-sac, less than sixty seconds into my Etosha-Africa experience, I'm a little off-balance, feeling that I've stumbled into some newly developing subdivision, in which the suburban tract houses have simply not yet been built, and the street—Zebra Avenue—not named yet.

Dennis hits the windshield washer, and the view smears quickly to the golden green of bug guts, a curtain of green, just beyond which the zebra graze.

We drive on to the main headquarters and register for our little cabin. There's a fenced-in area—a garrison, a compound—in which tents can be set up on a broad parking lot, theoretically safe from lions and leopards and elephants and other creatures; and within that central fort or garrison there is also a restaurant, gift shop, gas station, swimming pool, and so on. All vehicles must be out of the park, or back inside the garrison, by dusk each day, and are not allowed to go back out until dawn.

Like the day's specials on a menu, the times are posted on a marquee inside the park each night. There's a bit of a feeling of claustrophobia, as you realize you'll be hustled, as if with inmates, with everyone else in the park, after that witching hour, unable to leave, and it's not quite like anything else I've ever known.

Dennis and I have an hour and four minutes left before we have to be back inside the fort's gates—what happens after that, I do not know—and so we drive one of the suggested loops, stop-ping at the various observation stations, where we gaze through the windshield exactly as if attending a drive-in theater where the selected feature that evening is landscape.

Bands of ungulates drift across the chalky plains of the Eto-sha pan, the glaring brilliance of which is softened by the lowering sun. We watch wildebeests, with their powerful necks and shoulders and their immense heads, and the innumerable springbok, seeming as light and movement-driven as tumbleweeds, and more herds of zebras.

There is not the concentrated epic of seasonal migration. The animals have dispersed, with water briefly available at more locations than will exist later in the year, but still the assemblage is impressive; almost anywhere you look, in any direction, you can see something: a herd of oryx by the roadside, with another herd some few hundred yards beyond that, with another herd of something—springbok? zebra?—still farther back, and beyond that, a band of something else.

Waves of animals, each made smaller by distance, stretch to the flattened reach of the horizon, and all are in motion at this time of day, so that the general overall feeling is that the landscape is yielding a current, a jet stream, of larger purpose and desire in which the individual animals and herds are only embedded. It is as if there is a Coriolis effect of grazing animals out on that landscape, with their movements really no more free-willed than that which might be allowed by the geomagnetic yearnings or arrangements of certain electrons likewise entrapped in the fire-hardened stone far beneath them, calling out to the iron in their bright red blood.

They are all tacking somewhat in the same direction, all on the same march into the dry wind, and there is very much the feeling that after Dennis and I have gone back into the fort tonight they will continue marching, will keep on marching right off the map, and the edge of the world, and that others will take their place in the morning, similarly marching.

At one observation point, Dennis and I stare out at the bone-leached stretch of the Etosha pan, salt crystal white to the hori-

zon, with only the faintest tufts of grass growing here and there
—some animals travel around its edges, nibbling at the sparse
grasses, while others trudge straight across the barren saltpan, as
if food no longer matters to them. And viewing the shimmering
white horizon fifty miles on a surface flat as a dime, it's easy to
see what the German military was thinking: why they built a fort
here, where any force invading from the north would be visible
from a very long distance. What it was the Germans might have
thought they were protecting, amid all this heat and brilliance
and salt wind, is unclear to me.

Other rental cars toodle past, and VW camper vans, each of us
similarly caged and separated from the landscape. I still have such
trouble believing this is where we came from—such a fiercely
hot and waterless land, and with our puny lives made so available
to the meat eaters that have always followed these great ungulate
herds.

What would be our chances of survival today, Dennis and I
wonder, if we stepped out of the rental car, struck out across the
pan walking alone and upright, seeking to reach, say, Botswana,
and sought to arrive there unscathed, uneaten?

They seem slim indeed. In a landscape filled with six-hun-
dred-pound lions that are adept at pulling down six-hundred-
pound oryx or wildebeests, and in a land where the spotted leopards can climb a tree far better than any man, and certainly far
better than Dennis or I, it seems improbable that we could survive even a single night out on the pan by any means other than
chance luck or grace; and to consider a week or month of such
survival, a month or year or lifetime, much less a century or mil-
lennium or eon, seems as strange and impossible as any of the
other wonders we have seen on this trip.

We sit there in the rented jeep and roll our windows down a little farther to feel better the salt winds stirring across us. The day is becoming gentle, shadows are appearing upon the pan, the sighted world is vanishing from us, and soon the stars will be out. We glance at our watches, calibrating the necessity of our return. It's time to go.

Back in the compound, strangeness and surrealism rule; though why shouldn't it, when a species such as our own is corralled and penned into a constricted space, where our tendency to do what we do best—consume—is accentuated, and where another filament in our spindled character, a general uneasiness with and persistent clumsiness of our estranged relationship with the rest of the natural world, is also aggravated.

Everywhere, folks are grilling meat: burning the hell out of hot dogs and charring hamburgers on flames that crackle several feet above the grills. Every cabin, every bungalow, it seems, is cooking away, so that the desert evening is filled with the rank odor of scorched and burning meat, with the heavier, denser odors of lighter fluid thick throughout the compound, as if an unseen pool of it covers the ground: and if I were a predator, I don't know if I would be lured by the scent of all the burning meat or repelled by the razor-sharp petrochemical scent of so much lighter fluid.

Everywhere, flames are leaping from the back porches, and how none of the cabins catches on fire, I have no idea. The scent of gin floats heavy in the air too, currents and eddies of it pooling chest-high, so that there is a kind of olfactory sandwich going on, a lighter fluid–juniper berry–burned meat combination: and through such ritual, the voyagers stand uneasy guard over the otherwise uncontrollable interface between the daylight they are leaving and the deepening approach of night.

As Dennis and I walk through the compound, the entire village appears to be in flames.

There is another spectacle at Etosha. There is a pit, an arena, on the northwest end of the compound, where the animals come in to water each evening. The watering hole is illuminated with halogen intensity, and moths the size of small birds or bats swarm the lights, plummet to the ground. Visitors can go stand at a four-foot high rock wall and watch as the animals, blinking in the headlight-brilliance, come shuffling in to drink. There is even a little viewing grotto set up at one end of the watering hole, with a set of bleachers.

It's fascinating, part zoo and part wild. The animals must have their water, and have evidently accommodated themselves to the humans' unseen presence, scented but unable to be viewed behind that screen or curtain of fierce white light, and it's definitely not the nature to which I've become accustomed, solitary or almost solitary. It is a herd experience, people from all nations and of all ages standing shoulder to shoulder, whispering and conversing as the giants come striding down the sandy pit to drink, emerging from the darkness and into the light like fashion models on a runway.

The viewers are themselves eating and drinking as their eyes take in the watering hole's revelations, the images, and with only that puny wall separating us. At the base of the wall, on the other side, there is a jumble of rocks and some loose rope woven here and there, but it's nothing a lion couldn't leap without even trying, were the lion motivated to try to pierce or otherwise penetrate that shining fierce wall of light.

Skittish hyenas, terrifying to behold, immense and muscular and red-eyed, dart in to sip, then slink back into the darkness.

One of the showstoppers, a giraffe, strolls down to the water, braces its legs wide, then lowers that incredible neck to the pond's flat chalky waters and begins to drink, sucking the water up that long pipe. The giraffe drinks without ceasing for perhaps five minutes, maybe longer; it's possible to imagine, as in a cartoon, the water level in the pond dropping, maybe even going dry.

Sandpipers wheel and sip, skittering along the sandy shore, and after the giraffe leaves, the rhinos come, appearing one at a time, and entering from various directions: social, having arrived as a party, and yet testy or territorial or peevish or *something*. There's an electricity between them, a communicativeness and connection that I cannot discern, that none of us can discern, and yet that is present if unseen, as real between them as if tethered to one another with heavy ropes and harnesses. And whereas we have previously had to hotfoot it all over the stony desert, following tracks and the spoor of *Euphorbia* dung logs, we are now able to rest against the rock wall and stare down into a virtual *pit* of rhinos, wallowing and wrestling and horn-thrusting, pawing the sand and making mock charges at one another, squealing and roaring.

Are they real? Or are they clones or animated robots, skin-wrapped constructs of rib and spar, shuffling through a remote-controlled choreography designed to delight us, if even only for a little while, before attention spans lapse and the drink glasses are drained?

I try to gauge the age and physical condition of each rhino. There are six of them now, all different sizes and conditions. I wish Mike were here to tell me what I'm seeing, and I'm reminded, strangely, of Degas's painting *The Bathers*.

Indeed, what appears to be the oldest one—matriarch? patriarch?—wades resolutely out into the center of the pond and then just rests there, staking claim and luxuriating his or her old bones in the soaking warmth, the respite from the fevered day; and using, it seems, some of the water's buoyancy to ease the strain of holding his or her old bulk aloft all day.

And around this old boy or old girl, the other rhinos wander the banks like stars orbiting his or her perimeter, still earning and feeling their way in the world.

There is another mature rhino, a very large animal, though younger and more spry than the old pool-soaking rhino. She is attended by a very large calf whose antics seem to *reek* of boyishness. Her boy calf, Junior—as I'll call him—is being bullied by another, larger lone rhino, a young male—Bad Boy—who, all alone in the world for the first time perhaps, is, in time-honored fashion, truculently staking out territory, *any* territory: literally huffing and puffing and drawing lines in the sand, tossing his head, waving his horn about. Bad Boy is about 20 percent larger than Junior, but he is still an adolescent, or just barely post-adolescent, and Junior is the only rhino near his size.

Bad Boy promptly sets about bullying Junior, lowering his head and stalking, so that the junior rhino has to keep yielding, giving up space, tiptoeing away from the aggressive approaches of Bad Boy, who celebrates each new victory of useless sand gained with a triumphant thrust of his horn, and a little dance step.

These two, at least, are not here for the water. It seems that they are all here to simply hang out, to socialize, in the best way they know how.

Except for the two knuckleheaded young boys, it seems that the rest of them are just hanging around waiting for something to happen.

What do I know? Perhaps a lion is at large, herding them together; perhaps the stilted waiting quality attending them is instead tension, with the lion, or lions, just beyond the edge of darkness. I know nothing, am only watching it all for the first time.

Again, like clockwork or gearwork, the five rhinos station themselves around the circular pool like points of the zodiac; and though they are all canted toward the water, pointing toward it as a flower leans toward light, they make no move to travel the last few steps to reach it.

Perhaps it has something to do with the old sultan in the pond's center. Perhaps it has something to do with hierarchy: maybe so-and-so has to drink first. Or maybe they're simply not thirsty. They stand before us, seeming ill at ease—and who would not be, beneath the UFO brilliance of the overhead staging lights?—and continue to exist in their own galaxy, the one in which almost no one—maybe not even those who have grown up side by side with them, like Mike, or those who have hunted them, like certain of the guards who now protect them—can know what they are thinking.

And yet standing there at the wall with maybe thirty other observers strung up and down that low stone barricade, the throng of us pressed there as if standing at the edge of a park pond, into which we wait to toss bread crumbs for carp, or puddle ducks—I think that we begin to understand some basic things. It is as if loosened puzzle pieces, or what we perceived to be loosened puzzle pieces but were perhaps still attached and connected, ordered and aligned all along, reassert themselves or imprint upon us a perception of pattern or order, and we gradually learn things.

We come to see, for instance, Bad Boy is dying to stake out his own territory like the behemoths around him. The only one remotely his equal, however—Junior—a small toy of a rhino, who, we come to see and understand, is bored, as well as a little mischievous, with maybe also just the first few golden traces of hormone beginning to glitter and stir within his new but so ancient blood.

There can be no mistaking what Junior is up to as he sneaks out from behind the protective flank of his mother, charges Bad Boy, tossing his little head, his little horn, and acting like a tough guy, challenging the big street punk to a fight, only to then whirl and run back, trying to engage in the much-desired combat, or at least the much-desired turf battle.

Bad Boy is larger than Junior, and much stronger, and maybe a hundred times fiercer. He is all but strangling on his furies, and whether they are hormonal, or coming from some other rhino place, I have no way of knowing; but every time Junior runs up to Bad Boy, pretending to want to fight, Bad Boy blows a clot, drops his head, and lunges at Junior, who then has to whirl in the middle of his own bluff and run back to Mother, who pretends not to be watching any of this—who actually seems to be somehow choreographing it. Though every time Bad Boy is about to do harm to Junior, she intervenes at the very last second, taking one or two steps toward Bad Boy, which is always enough to make him slam on the brakes, almost falling down in his sudden change of heart, like a cartoon character now, while Junior all but struts and dances, safely back within the sphere of his big mother's influence.

We watch them for an hour and a half. Other people leave, drifting off for bed, until very few others are left. One little girl, about seven or eight, keeps looking in my direction and pointing and then whispering to her sisters. After a while she approaches, as shy as a shadow, and tells me that her mother works here at the park, that the girl herself is very familiar with the park, and that there is a snake that lives in the crevice between the very rocks against which I am leaning.

I step back quickly and thank her, and move several feet farther down the wall. Boomslang, black mamba, spitting cobra? Saved, by a child. Maybe it wasn't a poisonous snake, but then again, it was clearly one her mother had told her to watch out for.

I remember reading about recent lion maulings here, at this very wall: a lion leaping the piddly little low wall and killing and eating a couple of German tourists.

On and on, the boy rhino and the adolescent rhino play and fight, bait and switch, and the evening, like some beautifully irrelevant tiny cog in the gears of a century, eases forward with fly-wheel-fluttering frailty: such tiny movements conspiring to be part of such a greater and more ancient thing. I could not name the thing, but we are all staring at it when we look down into that sand pit: watching it, seeing it, even if the name for it does not register, and neither is it even on the tip of our tongues.

Later, it seems that the sand pit has filled, as if with some saturation of anger or angst—and that while all the other rhinos have just been standing there, waiting in their armor-clad patience, Bad Boy's attitude has been filling the pit like rising waters, contagious; and now, as if in some slow ballet, various rhinos are pairing off and roaring at one another, doing mock charges and head fakes and sometimes even clacking their long horns together—not trying to gore one another, I don't think, but sparring.

The sound of their roaring, and the clacking sound of their horn fights, is like something I have never heard or known, deeper and more primal than even the roar of a bear or the squall of a mountain lion. The only thing I have to remotely compare it to in my experience is like something from a Spielberg dinosaur movie, pterodactyl roars, which fill the pit and echo across the compound—and neither Dennis nor I can really tell what's going on, we know only that tensions have risen, that maybe the adults now are showing the youngsters how to grapple and fight, how to demand and defend turf.

And then when one turns to leave, they all do, within a matter of minutes, unwinding from their various compass points around the pond and walking back off into the darkness, with even the Old Bather climbing out, dripping, and following them slowly into the black.

Whatever comes next, I do not know—hyena, wildebeest, desert fox—for we can no longer stay awake, and so we walk back to our cottage, where the air conditioner is humming. I set the alarm clock for three, so that I can get back up and go check out the pit again, like a fisherman running a trotline. And I fall instantly asleep and dream that I am surrounded by hundreds of hungry carnivores, like a lone piece of meat in a sandwich, with the four walls of the little cottage like but slices of bread, and that my job, and my desire, is simply to keep from being eaten.

It is not a frightening dream, or even a tense one; if anything, a curious wondering inhabits it. It feels as if I am thumbing through the scrapbook or family album of some deeper, further assemblage of persons whose history was previously unknown to me, but who are, it turns out, direct kinfolk, who simply lie deeper in some vault than I had ever known existed: deeper than deep, and yet family.

After the alarm clock sounds and I swing out of bed and slip into my sandals, and I step out into the strangely muggy night—after the basalt aridity of Damaraland, perhaps even the presence of one cloud, or one tourist's swimming pool, or one watering hole, is enough to register as humidity—I'm struck by how odd, how consumer-like, it feels to be wandering over to my sure-thing rendezvous with wildlife.

The three a.m. part of it reminds me a little bit of hunting season, but the sandals and shorts most certainly do not, and neither do all the cottage units, each with its little porch light, nor the parking lots filled with cars and trucks and buses; nor does the certainty, the knowledge, that when I walk up to that low wall and peer over into the bright lights of the sand pit and the watering hole, I am going to see something. I'm not sure what, but there will always be something: the desert is too arid, and the nighttime needs of the animals too vast; they must have water, and like rough clockwork, if not quite as precisely as the celestial schedulings, they will rotate through their slots according to hard-negotiated confluences of need and opportunity. *Something* will be there.

It's chilly, and I've left my jacket in the truck. If I am to stand at the wall for an hour or more I'll probably want it, and so I find our truck in the parking lot and press the automatic door-unlocker, which emits its unavoidable *beep!* whenever the lock shifts status. This, too, is not like hunting—a yelping car in the middle of the night, when stealth is desired—and the car's electronic beep is answered immediately by the mad laughing howl of a nearby hyena.

I'm assuming that the animal—the *beast*—is just on the other side of the chainlink fence that presses up against the parking lot, back in the seemingly impenetrable tangle of brush, which is also pressing against that fence, though the sound is so close that I suppose it could be coming from the next parking lot over. I very much want the hyena to be on the other side of the fence—I feel ridiculously suburban in my shorts and floppy sandals, with my beeping $40,000 car. *Please don't eat me, Mister Hyena.*

Still, it is a bracing shock to me, this collision between eat-your-ass wild—not feral, but *wild*—and twenty-first-century commercialism. This is not the coyote in the Los Angeles suburbs, or even the bear wandering across the golf course at night. This is prehistoric, Paleolithic carnivorousness in the parking lot, or one warm breath away from the parking lot. This is the overflow of wildness, with everything turned backwards from the way we usually perceive it, with man dominant and omnipotent, exerting lordly rule over most of the globe, while the rank and uncontrollable things stand alone and fragmented now, to be found in only the most remote corners of the world, like the ill-fitting pieces of a puzzle that has long ago been sold for pennies in some dusty garage sale, *liquidated*, with that handful of lost pieces having slipped behind the couch, or even through the cracks in the flooring.

Here, it is not this way. Here, it is we who are like the dog in the kennel, or elephants in a zoo, penned in our tiny chainlink enclosure while the rest of the larger world presses in.

Here the single beep of a car alarm is seen as an affront, and is answered immediately. *Tone it down in 6-C,* the hyena might well have been howling—*or else*. Or else what? *Or else I'll come through the fence and tear your lungs out?* Well, yes, maybe.

Confident however that the hyena *is* back in the brush, and on the other side of that fence, which serves at least as a psychological screen, I begin my stroll through the complex, taking a shortcut through the sand and grass lawns and curving sidewalks that make up the little tourist neighborhood, moving from miniature streetlamp to streetlamp. The stars above are those of African skies, but I am in a suburb, and I have not traveled twenty paces before I hear, loud and clear over the hum and groan of all the air conditioners, the exceedingly close and assertive roar of a lion: the movie-roar coughing, though sounding a hundred times louder, and a hundred times more imminent.

I've spent much of my life in the woods gauging how far away animals are from me: the gobble of an approaching turkey as he answers my call, or the bugle of a bull elk, likewise approaching, or the low grunt of a whitetail buck, only twenty or thirty yards out. I know that sounds are sometimes closer than you realize, and what it sounds like to me is that the lion is in the compound—more than half the distance in already. It sounds to me as if he's only two houses away, and the sound of it stops my approach as if I have walked into a wall: my body will not compel me to go any farther.

I stand there breathing quietly, and listen again. The roar is coming from the direction of the watering hole. In the far distance, I can see the all-night glare of the stadium lights, but the roar is much closer than that, is emanating from a point between me and there.

And yet this is not possible—this is ridiculous. A national park wouldn't let lions wander around inside, at night, would it?

Surely my instincts, as well as my experience, is mistaken, and the controls of the park and the government are accurate and adequate. It just *sounds* like the lion is closer, and my body, though refusing to take another step, is also mistaken. I stand there, breathing and listening, and again the roar sounds, though this time it sounds two houses farther on, as if the lion is prowling now, examining the barbecue grills, perhaps, going from patio to patio, investigating spilled crumbs of char, and licking the grease and fat from the grills.

Surely I am mistaken. A park is a controlled environment. A park is safe.

I force myself to walk slowly toward the bright lights, which are still a good fifty yards distant. The roaring has stopped—I am not sure what this means, and it occurs to me, in this strange juxtaposition of disobeying what my body is telling me, that I am moving as if in a trance: a disobedient sleepwalker, proceeding under the flimsy veneer of an illusion that all is well.

It does not seem to require too much imagination to envision a world otherwise—a world in which the desire and authority of man are not always paramount—and yet, under cover of darkness, and passing from one cast of cute streetlamp to another —avoiding the sidewalks, and cutting through the yards, and walking ridiculously along the arced perimeters of those casts of lamplight—it seems as if all that is required for me to remain safe is to stay at the edges of those shadows.

The lion roars once more. Again, it paralyzes me, sounding definitely between me and the sandpit (as if the sandpit itself, with its piddly four-foot wall, is safe!). And yet I have it in my mind that if I walk up to that wall and peer over, I will behold a giant African lion, standing there amid the bedazzle of light. Not drinking water, but just roaring, looking back at me, aware of my upwind scent and my beepy car alarm, and roaring.

My mind is trying to tell me it would be like watching a lion at a zoo—that this fellow (if it is a male) probably is used to letting people watch him, that he means nothing by all his roaring —but my body is telling me to not take one step farther, that my desire to see a lion is carrying me headlong into disaster.

I try to argue with my body. The consuming American part of me tries to tell my body and my instincts, *Lookit, have I not set the alarm clock to awaken at three so that I might walk to that wall and behold this very thing, a lion?* The shopper in me does not want to be denied.

There is no one else out. And because the lion has been so loud and so close, it is difficult to imagine that everyone is sleeping through such a din, and the impression I have is that instead everyone else is hiding, cowering, with the only sound in the compound now that of the united hum of two hundred or more air conditioners.

If only I had a spear, or anything! A plastic Wiffle ball bat, a pocketknife, a circus whip!

The lights are very close now—if only I can make myself propel myself through that last twenty yards of space, I will be there, will be able to peer over the wall and see that which has fallen silent, and which is waiting for me—and yet now, the closer I get, the harder it is to move.

I stop, as if I have been running a long way while carrying a heavy load—a backpack loaded with iron—and consider further this unlikely and unaccustomed dialogue between body and mind. Instinct versus desire.

When I was a child growing up in Texas, I once amused my

parents mightily while on a jungle-themed boat ride at a place called Six Flags Over Texas. The bank-side thickets were rustling as the murky channel upon which our little bamboo gunboat puttered grew ever narrower, and mechanical hippos and crocodiles boiled up from beneath the surface, the vile bayou water funneling from their snouts as they gasped and clacked their jaws. Wild-eyed savages lunged from the thickets, their faces painted wildly like cannibals! And while some of the giant creatures were clearly mechanized, this was no less terrifying for me, for it seemed we had entered a world where even the mechanical creatures meant harm.

Acrid blue smoke drifted in thick clouds across our tapering channel—the ever-narrowing anti-birth canal—and the boat paused. Up ahead was a small cave, barely larger than our craft, and the brown waters roiled past us, with the river continuing on into that cave. Our captain, our guide, was a silly fellow in khaki shorts and epauletted bush shirt, wearing a pith helmet. He was young and ruddy, hale and beefy—perspiring, excited-looking. He had a megaphone, which he brandished before us, shouting to make himself heard over the wail and whistle of cannons.

"Do we push on," he inquired of us, his passengers, "or do we turn back?"

Imagine my horror, when like disembodied, soul-snatched zombies, every one of my fellow passengers began to answer, at first one by one but then in gathering unison and with glee, "Push on! Push on!"

This was not where I wanted to be, this was not the world I had signed up for! The passengers were laughing—stupid dipshits!—in a way that made me loathe them more than I had ever loathed anything: hating them with the anger of betrayal.

The dopey boat captain—radiating vapid foolhardiness, *beaming* at what he believed was his good fortune at being in command of so intrepid a bunch of passengers—addressed us one last time as the boat began to stir and then shudder, as the underwater pickup tracks engaged once more, preparing to ferry us forward to our doom.

"Are you sure?" he asked. "Is every one of you certain? Should we push on," he asked again, "or turn back?"

Previously petrified, I now rose from my seat—extricating myself from the pressed-flesh handbasket of all the other happy and clueless campers—and shouted, summoning from my deepest core, "Turn back! Turn *back!*"—roaring it with all the force and power I could gather, as seeking to negate, absorb, and counter the profoundly mistaken choices and direction of the twenty or more other passengers.

I recall there was another pause. For a moment I believed I had pulled it off—I had turned the *Titanic,* had impressed my fear upon each and every passenger—but then they burst into laughter, and the college kid—who, to his credit, had been briefly startled (wondering, I realize now, if the boat even had the ability to reverse course)—shook his head and said something along the lines of "The rest of the crew wants to go forward, matey." He dropped a lever, thunked the boat into gear, and, with the rest of the passengers still laughing, and even my parents smiling and assuring me that it would be all right, that we would make it out okay, we proceeded to continue on with our descent into dark hell.

Here in Etosha, forty years later, there are no passengers, but my body remains seized with fear as it did then. And while there is now a lazier or more jaded part of me that wants to view Etosha as merely another theme park, there is as then a deeper, older part that, right or wrong, disagrees with that easier and more secure interpretation.

I can't go any farther. I could force my mind to force my body to step up and behold the lion, which has fallen silent, and which I suspect would be looking right at me—but such forcing would be its own betrayal, and I can't do it. Fear is cold within me, though it is not the fear that has stopped me iron-tracked, but the other thing; to betray what I am feeling so deeply would be like stepping through a plate of glass and abusing the trust of my body as well.

I stand there, so close now—only another ten yards—and am so tempted.

A small brown animal stirs at the edges of my vision within the compound; it comes trotting toward me. It's a jackal, coming quickly—looking exactly like the coyotes of home, though a bit smaller—and, already rattled, I'm startled further, and raise my arms like a Hollywood ogre, and the jackal flares but does not run away, and I wonder if it is mad, or starving, to be so bold.

If I turned my back on it and walked away, would it trot along behind me, waiting to rush in and bite at my ankles, my hamstrings? Even as I am considering such, the jackal drifts in closer again, and when I make an aggressive step toward it, it skitters away, but only as far as my own advance carried.

Why is the lion not roaring anymore? And why is this jackal over the wall and at large in the compound, while the barbecuenation of men and women sleeps? Did the lion and the jackal come over together? At home, in my valley, coyotes have a close association with mountain lions, letting the lions do the killing and then scavenging the remains in the days following.

It occurs to me that I am a long way from my cabin: that even though it is but another ten yards to the wall, I have come too far. What if I have *passed* the lion and it now lies between me and my return?

The little jackal is feinting toward me again, and once more I shoo it away, though there continues to be a pattern of diminishing return on such action, so that soon, it seems, the jackal will be hanging from my hand or wrist.

The lion is waiting there, listening and watching and waiting for me to peer over the wall: there is too much tension and stillness, as opposed to more quietness. I turn and walk carefully, heart thumping, back toward my cabin. The little jackal trots along behind me as if it is my dog.

I reverse my path, walking now from streetlamp to streetlamp. It is too dark, out in the raked sand, to discern any tracks. In the morning, I imagine, groundskeepers might rake the sand smooth anew.

I make it back to the cabin safely, where I step into the chilly room, close the door—one more unit among two hundred—and slip out of my sandals and climb into bed, beneath the warm covers, where I feel deliciously alive, and am delighted by and marvel at this simple fact.

In the morning we have time for only a brief spin through the park, a couple or three hours' worth, which, when you are accustomed to being able to get out and walk, is really long enough; plenty of car time, if not too much. We swing by the watering hole, curious—gone are the megafauna of the night before, as well as everything else. There is only a smallish pool of muddy brown water surrounded by sand; nothing about the place seems to suggest even a hint of the fantastic procession of events of the night before. Under the bright heat of the day, the arena, the sand pit, seems to be utterly lacking in the ability to spur or ignite even the imagination for such things as we witnessed there the previous evening. As if the world had been turned upside down in darkness.

Out on the dazzling Etosha pan, the same clocklike procession continues the windswept drift, the animals with their heads down and tails swishing. There is always something, some ceaseless life, migrating across the vast emptiness, and though it is a child's question, unanswerable, it arises more insistently than ever, more glaringly than ever: What is the difference between lifeless rock and a wildebeest, or between wind and a zebra, and why are there wildebeests and zebras instead of just stone and heat and wind? I can understand the former, and I can even understand (though barely) the atomic rearrangement of things that results in the brief phenomenon of life.

But I still cannot understand the stupendous specificity of the wildebeest's stripes, the wildebeest's hooves, the zebra's bands, or the need or reason for the numbers of them flowing endlessly, as if haunted and hungry, searching for something more than just grass — owned, possessed, directed, *driven*. Even a rudimentary understanding of probability leads to the realization that the math doesn't match up: that no matter how tightly you squeeze a chunk of basalt, you should not be able to get zebras and giraffes to emerge from it, much less rhinos — or, for that matter, individuals such as Mike Hearn.

Forty-six years, once more I am tongue-tied, amazed and astounded.

We cruise through the shimmering heat like astronauts in a moon buggy. The road winds through little thickets of brush in which brightly colored birds flutter and perch, and, as if consulting a foreign-language dictionary, I try to identify these scraps of song, these brilliant flags of color shedding, for now, the unanswerable questions of probability, and trying, like an infant, only to learn their names—or rather, the names that we have given them. *Melodious lark, monotonous lark, pale chanting goshawk. Sunbird, sugarbird, wailing cisticola.* As with our own continent's species, a few are in need—as noted by the poet Jim Harrison—of renaming: the *brown robin,* for instance, or the *familiar chat*—though all in all, the African names seem superior: the white-rumped babbler, the black-faced babbler, and the kurrichane thrush. We spy a Fülleborn's longclaw scuttling into the shade of a little grove, its yellow and black plumage painted so carefully by the hands of time, but also, clearly by something else—though *what,* no one yet has been able to prove or identify, even as the presence of that thing, that divine imagination or divine force, or divine luck, surrounds us everywhere, anywhere, we care to look.

By this time tomorrow, we will be heading home, climbing onto a plane for the beginning of forty-some hours' worth of flight. We stop at a little camper's outpost on the far southeast corner of the park, an uninhabited tourist village; the handful of visitors who've come here in the height of African summer are all out in the park already, driving around. It's late morning, and we eat a silent breakfast in a dark and deserted restaurant, where we each reside quietly before our thoughts of home: a wall of such thoughts before us now not like a malaise of homesickness, but the sweet, dreaming in-between time of getting there but not quite yet arriving there, with the trip ending, even though it has not yet really ended.

The restaurant is cavernous, and despite its abandoned hush, it's easy to imagine it filled in busier times with seething throngs. As it is, Dennis and I are attended to by a waitstaff of six, and four chefs. Ivory tusks line the walls, and we sip coffee for a good half-hour after we've eaten, sunk into ceiling-fan somnolence. It's strange to remember that it's going to be Christmas—*is* Christmas—when we get back. Malls, singing electronic teddy bears, and the whole heartbreaking wrong-turned shitaree.

Back out into the sunlight then. We get in our truck and drive slowly, with waves of ungulates parting before us. It's a treat and a privilege to be here, and to be seeing such things, but how much finer it was to be following 275 yards behind the rhinos, or even to be out wandering across the rhinos' basalt, with no other life in sight, but out there, somewhere out there, indisputably known, if not yet seen. A thing that was once almost lost, now saved, and still with us.

There are strange days when it seems that we are so new in the world as to be ridiculously helpless and hopeless, and other days when the opposite feels true: that already, despite our incredible newness—not even two hundred thousand years under our belt—it feels as if we have been here too long.

As if we exist in a seam, along a fracture line of some sort, where one side is joy and the other is loneliness. And that even here, amid all of Africa's bounty, it is possible to skip across that boundary, anywhere, at any given time.

And who or what made, or designed, or negotiated such a boundary, and why?

Too many questions. There are still countless facts to be memorized, then learned, before wrestling, really wrestling, with such questions. And was that really a Fülleborn's longclaw we spied ducking into the brush, or might it have been another type of longclaw?

I went to Namibia wanting only to see something new—to step out of a rut, wonderful as that rut was, and is—and, despite having that wish delivered in spades, got lured into other things, considered the big questions. We didn't even get to touch the rest of Namibia, the red clay hills of the Himba, or the rock art of the Brandberg Massif, or the borderland, the beginning, of the Kalihari, or the great sand dunes above the Pacific, or the Skeleton Coast itself. We visited only the one fountain or birthplace of rhinos, Damaraland, and a wedge, a window, a shuttered, half-dreamed glimpse of Etosha—and yet the air, the milieu, was a matrix of *Why?* for all the big questions: brainstormers, and nonstop stimulation that was often simultaneously refreshing and exhausting, and ultimately addictive, so that even as we were leaving, we were planning our return: imagining the next places we would go, and of how we would take Mike up on his invitation to show us more cracks and corners of the wild country, wild continent, that he had made his new home. His home, period.

And then we were gone, holding tightly to the images, notes, memories and photos, and the artifacts—a fragment of stone, a twist of desert wood—and knowing from experience that despite the wondrousness of everything, all but the brightest memories would begin to fade almost immediately, destined to lose their luster and potency, their magic, in the other-world, the after-world, that lies always beyond the experience, always beyond the moment. The dim future, which can also be wonderful, but can never be quite the moment again itself.

Mike Hearn died ten days later. Dennis and I are back in Christmas-land, with our memories and emotions still as fresh as vital as a new-caught fish, still gleaming and vibrant, not yet beginning to lose any of that magical luster. He had finished up with the students and was taking his two-week break, his R-and-R in-country tour on which, lacking anyplace else to live in Namibia save his thatched hut tree house at Palmwag, he was renting a little closet of a cottage and hanging out on the windy, chilly coast, running and eating fresh seafood and surfing, riding mountain bikes up into the craggy black mountains that towered above the ocean in the heat of the day, and then surfing again. Drinking a cold beer in the evenings and visiting with old friends, and always making new ones.

Treading water, waiting to get back to his beloved rhinos and his beloved bedrock basalt country, with or without a life partner. *Thinking* about settling down: but not yet, not really. Thinking more about moving some rhinos, expanding their range; thinking about releasing them back farther into the world, as a designer might once first have considered stirring ash and clay and water to form and then distribute other dreams, other desires.

What Dennis and I heard was that he had an epileptic seizure while surfing. We had no idea whatsoever that such a condition, such vulnerability, existed anywhere within him. But it did, and it struck him while he was out surfing, and he drowned, gone like candle blink, with all who knew him disbelieving that one of the finest people in their lives—one of the best hearts—could be gone. No man or woman is stronger than the entirety of an ocean, but even those of us who knew him only peripherally thought of him as more elemental than a regular person: we thought of him, I think, as a desert, or as the rhino-man, or as a pure heart elementally distilled, irreducible, and as such, invulnerable in the density of that reduction; as close to perfect as we ourselves, were we so empowered, might dream a man or a woman.

And who had dreamed him? Some farther dreamer now or then? Mike's parents, partly at least, surely, and, no doubt, Mike himself, living and participating in his dream, each day. Building, dreaming, the man of himself.

We sent notes to his family in England but were unable to attend the various ceremonies and celebrations held in his honor.
I sent a note to his parents telling them how he had commented
how grateful he was, how remarkable he found it, that they had
supported him in this adventure—though *adventure* was not the
word he used, viewing all of it not as anything bold or unusual,
but merely an engaged hunger unscrolling one day after another,
with each day an opportunity to behave nobly: to exercise the joy
and power of presenting that heart to the world, and knowing
the pleasure of fit—that the world, for a while, would accept and
even embrace such a heart. And then he was gone, though there
were, and still are, nearly a couple of thousand rhinos left, out on
the volcanic black and red scablands of where, briefly, he was.

Being a hunter, for the longest time I used to think that there
might exist somewhere in the world a crack of secret knowledge,
that there might be a fissure in time somewhere in which proof
of certain things might be revealed, proof of what we suspect in-
tuitively, or by some other means of knowing, might be glimpsed,
as if looking down into some secret, mist-shrouded valley. What
is the difference between a rock and a rhino, or the wind and a
zebra, and what hand dreams and divines and then separates and
then reconjoins such things? Who came up with the whole ashes-
to-dust gig, in which even a zebra, or even a man having emerged
from basalt, or some other stone, returns to such—vanishing?
And the bigger question, amid such a system: Why are we, the pale
or fragile ones, here, and what, if anything, are our obligations?

I still think that it might be that way, that one might yet
glimpse—not so much by searching, but perhaps by happenstance, by not searching—in a single moment that truth or understanding, of the relationship of design, destiny, of plan and

dream and scheme versus mere chance. That one might finally know with certainty if the world is made and dreamed, or merely assembled by nothingness. I remain open to that possibility, while understanding also that we have a tendency or disposition as a species to see what we want to see, and to sometimes avoid seeing that which we do not want to see.

Like a hunter, then, I still sometimes wonder, with occasional frustration, where that last grain of science might reside that would reveal such knowledge—the blueprint, the secret map in the basement—and I wonder what bargain or trap might be set to lure or attain such knowledge, which for so long has evaded and avoided all other men and women.

The answer seems obvious—the only lure we have to offer the force that lies on the other side of that distance, the only currency of exchange, the only item of interest with which to barter, is our life itself, with the living of it somehow important, all the way up to that final transaction, and the beginning, then, of that initial understanding that we cannot know until then but can only prepare for: building and living a life of worth so as to present our lure, our bait, in such a manner as to perhaps draw the attention and interest of that previously unseen if not unsuspected force.

I remember the little sounds Tina was making when she came up the hill to where we were hiding, her ribs heaving not from the effort, but from her agitation: her suspicion that she was being watched.

The world seemed different, with her so close. It seemed that we had somehow found that secret cleft, that hidden passage through the mountains, or at least that notch in the mountain wall, through which we could peer at the other and equally wonderful, or perhaps even more wonderful, valley lying below.

And it seemed that some force from that other valley had alerted her to the fact that intruders were very near now, that they were peering through that notch, into a land they could not yet reach, were not yet allowed to enter.

Who else could have told her; how else could she have known?

I remember how she stood there poised above us, with her horn of God, the twin horns, canted slightly in our direction. I had never seen an animal so beautiful. We were looking right at that answer, the one that cannot otherwise be answered, and yet still we could not see it: though we could see her, by God, and she was something.

And how must we have appeared to her, at that point-blank range? Vague shapes and outlines and hues of slightly different earth tones? And what odd scent, odd clue, must have wrinkled itself across the near-zero space between us? After fifty-five million years of absence, must she too have divined that she was looking across at something significant beyond her old order of things—as if she herself had gone backwards across that distance, back toward a deeper time, closer, perhaps, to the time of her original making?

For whatever reasons, she stood there panting, then groaned, and turned and ran. She ran without stopping, galloping, toward the horizon, raising plumes of red dust as if each strike of her feet was kindling a fire, and with the little cream-colored calf, Ongoody, hurrying to stay with her: both of them racing away as if into the future, leaving us stranded in the present, wondering at what we had glimpsed, a little frightened and a little rapturous, and unwilling—unable—to turn away from the possibility of maybe one day seeing something like it yet again.

Acknowledgments

It seems strange that to encounter an individual who loves life as much as Mike Hearn did should be unusual, even uncommon. Strange too is this rarity in which a man or woman loves a place with such wholehearted enthusiasm—and is aware, in every incandescent hour, of his or her great fortune to have discovered and then been welcomed into that place. I'm grateful that I got to meet him, and feel robbed that less than a month later he was gone. His absence leaves a void in the lives of all he touched, but left behind an incredible conservation ethic in a newly independent Namibia, and for a species, the black rhino. Mike Hearn's passion for these things, and the example of his life, may well yet be part of what helps that improbable and seemingly ill-shapen creature pass safely through the eye of the needle of this, the Anthropocene—the greatest wave of species extinction since the Cretaceous die-offs.

I'm grateful to Round River Conservation Studies' Dennis Sizemore for introducing me to this landscape, and to the Round River students and staff, particularly the naturalists Jerry Scoville, Doug Milek, Chris Filardi, Suzie Dain-Owens, Chris Lockhart, Jeff Muntifering, and Kim Heinemeyer, as well as the mapping specialist Rick Tingey, and all the rest of the RRCS staff. In Namibia, I'm grateful for the guidance and hospitality of Chris Bakkus and Emce Verwey, and Simpson Uri-Gop, and to my family,

ACKNOWLEDGMENTS

who has traveled to Damaraland with me on various occasions. As ever, I'm deeply grateful for the assistance of my editor, Nicole Angeloro, and for line editing from Alison Kerr Miller and production editing from Lisa Glover, and to my agent, Bob Dattila. Megan Ruggiero helped with the preparation of the manuscript, for which I'm grateful, and I appreciate also the art and book design by Patrick Barry and Melissa Lotfy.

Portions of this manuscript have appeared, in slightly different form, in *Tricycle: The Buddhist Review* and *OnEarth* magazine; I'm grateful to the editors of those publications.

While Mike Hearn's life was singular, his efforts, fortunately, were not. By no means exhaustive, a very short list of other activists who have labored long on behalf of the black rhino includes Duncan Gilchrist, Garth Owen-Smith, and Rudy and Blythe Loutit, founders of the Save the Rhino Trust. Blythe died in 2005 after a long illness, a few weeks after Mike's passing. The contrast between the impermanence of any activist and the landscape he or she seeks to protect is always profound, but it is hard to imagine a broader gulf than the one that lies between the unchanging Namib Desert, static for tens of millions of years, and the quick flame of Mike's own short years. That he was so successful in his brief time is testament almost exclusively to the depth and breadth of his great and good heart. He will be missed for as long as we are here.